Methods and Protocols in Food Science

Series Editor
Anderson S. Sant'Ana
University of Campinas
Campinas, Brazil

For further volumes:
http://www.springer.com/series/16556

Methods and Protocols in Food Science series is devoted to the publication of research protocols and methodologies in all fields of food science.

Volumes and chapters will be organized by field and presented in such way that the readers will be able to reproduce the experiments in a step-by-step style. Each protocol will be characterized by a brief introductory section, followed by a short aims section, in which the precise purpose of the protocol will be clarified.

Emerging Food Processing Technologies

Edited by

Mohsen Gavahian

Department of Food Science, National Pingtung University of Science and Technology, Pingtung, Taiwan

Humana Press

Editor
Mohsen Gavahian
Department of Food Science
National Pingtung University of Science and Technology
Pingtung, Taiwan

ISSN 2662-950X ISSN 2662-9518 (electronic)
Methods and Protocols in Food Science
ISBN 978-1-0716-2135-6 ISBN 978-1-0716-2136-3 (eBook)
https://doi.org/10.1007/978-1-0716-2136-3

This Humana imprint is published by the registered company Springer Science+Business Media, LLC part of Springer Nature.
The registered company address is: 1 New York Plaza, New York, NY 10004, U.S.A.

Series Preface

The *Methods and Protocols in Food Science* series is devoted to the publication of research protocols and methodologies in all fields of food science. The series is unique as it includes protocols developed, validated, and used by food and related scientists as well as theoretical bases are provided for each protocol. Aspects related to improvements, adaptations, and further developments in the protocols may also be approached.

The *Methods and Protocols in Food Science* series aims to bring the most recent developments in research protocols in the field as well as very well-established methods. As such the series targets undergraduate, graduate, and researchers in the field of food science and correlated areas. The protocols documented in the series will be highly useful for scientific inquiries in the field of food sciences, presented in such a way that the reader will be able to reproduce the experiments in a step-by-step style.

Each protocol will be characterized by a brief introductory section, followed by a short aims section, in which the precise purpose of the protocol is clarified. Then, an in-depth list of materials and reagents required for employing the protocol is presented, followed by comprehensive and step-by-step procedures on how to perform that experiment. The next section brings the dos and don'ts when carrying out the protocol, followed by the main pitfalls faced and how to troubleshoot them. Finally, template results will be presented and their meaning/conclusions addressed.

The *Methods and Protocols in Food Science* series will fill an important gap, addressing a common complaint of food scientists, regarding the difficulties in repeating experiments detailed in scientific papers. With this, the series has a potential to become a reference material in food science laboratories of research centers and universities throughout the world.

Campinas, Brazil ***Anderson S. Sant'Ana***

Preface

The food processing industry is a dynamic and competitive area. Nowadays, consumers have a better understanding of food products and are looking for high-quality, safe, and convenient foods that are produced in a sustainable approach. The safety and quality parameters of the food (e.g., nutritional value, taste, and texture) are affected by the technique that the food industry utilizes for food processing. Many of the food factories rely on traditional/conventional food processing technologies due to simplicity and availability of these technologies. However, challenges associated with traditional food processing technologies, in many cases, cannot meet the consumers' demands. Hence, the food industry is looking for alternative processing technologies to address the limitations associated with conventional food processing techniques. This industrial need opens new areas of study to researchers.

Hence, several novel food processing technologies have been proposed in the past decades, and researchers around the world are working hard to better understand the sustainability, benefits, efficiency, and potential applications of these emerging technologies. These studies introduced new concepts and parameters that can affect process efficiency and product quality. Precise control and measurement of these parameters, which are related to emerging food processing technologies, is a crucial consideration for understanding the concepts and designing appropriate and optimized food processing technologies. Therefore, there is a need for protocols and procedures related to emerging food processing technologies.

This volume of the book aims to introduce procedures related to measuring the process parameters involved in some of the most important emerging food processing technologies and the approaches to measure the process efficiency. It covers both thermal and non-thermal emerging food processing technologies.

This book complies with the general standards of *Methods and Protocols in Food Science* (MeFS) series and will follow the general format of the MeFS. All the chapters of this volume are designed to contain an introduction to an emerging food processing technology, a list of the equipment and materials required for the analysis of process parameters and process efficiency along with clear, reproducible, and simple step-by-step protocols. Besides, each chapter contains considerable amounts of valuable notes to provide hints that can prevent mistakes and suggest troubleshooting approaches.

Hence, this book is the first book in this area of science that provides a comprehensive source of well-established procedures and protocols that can be used for analyzing the process parameters and efficiency of emerging processing technologies. The content of this book can be used in academia, research centers, and industry. This book will be particularly useful for young scholars, such as graduate students and postdoctoral associates, and food scientists who want to explore this novel and promising area of food processing. This book also provides fundamental procedures related to novel processing technologies for new researchers and students in this field. Therefore, it can be also used as a textbook/reference for teaching in academia, especially for students who are familiar with the concept of conventional food processing, in courses such as "emerging food processing technologies". Besides, the food industry and food machinery manufacturers may get benefited from the comprehensive information provided in this book.

This book is prepared in two sections. The first section (chapters 1 to 11) describes the procedures and methods that are used for measuring the process parameters and process efficiency of some nonthermal emerging food processing technologies. The second section (chapters 12 to 16) focuses on describing methods and procedures used for measuring the process parameters and efficiency of some thermal emerging food processing technologies. Both sections follow a similar format and provide fundamental information regarding the materials, tools, and methods needed to collect important data related to emerging food processing technology. The editor of this book, Dr. Mohsen Gavahian, would like to appreciate the series editor Prof. Anderson Sant'Ana for the kind invitation to edit this book. Also, the valuable contribution of authors to this volume of the book is highly appreciated. Finally, the support of publishing editor, Monica Suchy, and Springer Nature publishing team is highly valued.

Pingtung, Taiwan ***Mohsen Gavahian***

Contents

Contributors

ASAAD REHMAN AL-HILPHY • *Department of Food Science, College of Agriculture, University of Basrah, Basrah, Iraq*

KARIM ALLAF • *La Rochelle University, LaSIE UMR 7356 CNRS, La Rochelle, France*

TAMARA ALLAF • *ABCAR-DIC Process, Compiègne, France*

MANREET BHULLAR • *Department of Horticulture and Natural Sciences, Food Science Institute, Kansas State University, Olathe, KS, USA*

JOSÉ A. CANO-BUENDÍA • *Facultad de Medicina Veterinaria y Zootecnia, Department of Microbiology and Immunology, Universidad Nacional Autónoma de Mexico (UNAM), Mexico City, Mexico*

FARID CHEMAT • *INRAE, UMR 408, GREEN Extraction Team, Avignon University, Avignon, France*

MAGDALENA DADAN • *Department of Food Engineering and Process Management, Institute of Food Sciences, Warsaw University of Life Sciences – SGGW, Warsaw, Poland*

KATARYNE ÁRABE RIMÁ DE OLIVEIRA • *Laboratory of Food Microbiology, Department of Nutrition, Health Sciences Center, Federal University of Paraíba, João Pessoa, Paraíba, Brazil*

LOUISE IARA GOMES DE OLIVEIRA • *Laboratory of Microbial Processes in Foods, Department of Food Engineering, Center of Technology, Federal University of Paraíba, João Pessoa, Paraíba, Brazil*

JEFFERSON DE OLIVEIRA MALLIA • *Department of Food Sciences and Nutrition, Faculty of Health Sciences, University of Malta, Msida, Malta*

EVANDRO LEITE DE SOUZA • *Laboratory of Food Microbiology, Department of Nutrition, Health Sciences Center, Federal University of Paraíba, João Pessoa, Paraíba, Brazil*

CARLOS ESCOTT • *enotecUPM (Enología, Enotecnia y Biotecnología Enológica). Chemistry and Food Technology Department, ETSIAAB, Universidad Politécnica de Madrid (UPM), Madrid, Spain*

MOHSEN GAVAHIAN • *Department of Food Science, National Pingtung University of Science and Technology, Pingtung, Taiwan*

STEPHEN GITERU • *AgResearch, Palmerston North, New Zealand*

S. M. A. ISLAM • *Department of Animal Science, Bangladesh Agricultural University, Mymensingh, Bangladesh*

PITIYA KAMONPATANA • *Department of Food Science and Technology, Faculty of Agro-Industry, Kasetsart University, Bangkok, Thailand*

ANJINEYULU KOTHAKOTA • *Agro-Processing & Technology Division, CSIR-National Institute for Interdisciplinary Science and Technology (NIIST), Trivandrum, Kerala, India*

SZE YING LEONG • *Department of Food Science, University of Otago, Dunedin, New Zealand; Riddet Institute, Palmerston North, New Zealand*

JENSHINN LIN • *Department of Food Science, National Pingtung University of Science and Technology, Pingtung, Taiwan*

IRIS LOIRA • *enotecUPM (Enología, Enotecnia y Biotecnología Enológica). Chemistry and Food Technology Department, ETSIAAB, Universidad Politécnica de Madrid (UPM), Madrid, Spain*

MARCIANE MAGNANI • *Laboratory of Microbial Processes in Foods, Department of Food Engineering, Center of Technology, Federal University of Paraíba, João Pessoa, Paraíba, Brazil*
GIRISH N. MATHAD • *Department of Tropical Agriculture and International Cooperation, National Pingtung University of Science and Technology, Pingtung, Taiwan*
N. N. MISRA • *Department of Engineering, Faculty of Agriculture, Dalhousie University, Halifax, NS, Canada; Ingenium Naturae Private Limited, Bharuch, Gujarat, India*
ANTONIO MORATA • *enotecUPM (Enología, Enotecnia y Biotecnología Enológica). Chemistry and Food Technology Department, ETSIAAB, Universidad Politécnica de Madrid (UPM), Madrid, Spain*
H. M. MURSHED • *Department of Animal Science, Bangladesh Agricultural University, Mymensingh, Bangladesh*
JOELLE NADER • *Lebanese American University, Byblos, Lebanon*
MINH KHOA NGUYEN • *Department of Food Science, National Pingtung University of Science and Technology, Pingtung, Taiwan*
MAŁGORZATA NOWACKA • *Department of Food Engineering and Process Management, Institute of Food Sciences, Warsaw University of Life Sciences – SGGW, Warsaw, Poland*
INDRAWATI OEY • *Department of Food Science, University of Otago, Dunedin, New Zealand; Riddet Institute, Palmerston North, New Zealand*
SHIKHA OJHA • *Quality and Safety of Food and Feed, Leibniz Institute for Agricultural Engineering and Bioeconomy (ATB), Potsdam, Germany*
R. PANDISELVAM • *Physiology, Biochemistry and Post-Harvest Technology Division, ICAR-Central Plantation Crops Research Institute (CPCRI), Kasaragod, Kerala, India*
CONRAD O. PERERA • *School of Chemical Sciences, Food Science Program, The University of Auckland, Auckland, New Zealand*
SANDRINE PERINO • *INRAE, UMR 408, GREEN Extraction Team, Avignon University, Avignon, France*
EMMANUEL PETITCOLAS • *INRAE, UMR 408, GREEN Extraction Team, Avignon University, Avignon, France*
YUN PING NEO • *School of Biosciences, Taylor's University, Subang Jaya, Selangor Darul Ehsan, Malaysia*
KRISHNA PRABHA • *Department of Chemical Engineering (Food Technology), Vignan's Foundation for Science, Technology & Research (Deemed to be University), Guntur, India*
V. PRITHVIRAJ • *Department of Food Engineering, National Institute of Food Technology Entrepreneurship and Management, Sonipat, Haryana, India*
S. M. E. RAHMAN • *Department of Animal Science, Bangladesh Agricultural University, Mymensingh, Bangladesh*
JUAN C. RAMÍREZ-OREJEL • *Facultad de Medicina Veterinaria y Zootecnia, Department of Animal Nutrition and Biochemistry, Universidad Nacional Autónoma de México (UNAM), Mexico City, Mexico*
AMANDINE ROUSSET • *INRAE, UMR 408, GREEN Extraction Team, Avignon University, Avignon, France*
SUDHIR K. SASTRY • *Department of Food, Agricultural and Biological Engineering, The Ohio State University, Columbus, OH, USA*
OLIVER K. SCHLÜTER • *Quality and Safety of Food and Feed, Leibniz Institute for Agricultural Engineering and Bioeconomy (ATB), Potsdam, Germany; Department of Agricultural and Food Sciences, University of Bologna, Cesena, Italy*

DAVID SPITERI • *Department of Food Sciences and Nutrition, Faculty of Health Sciences, University of Malta, Msida, Malta*
VASILIS VALDRAMIDIS • *Department of Food Sciences and Nutrition, Faculty of Health Sciences, University of Malta, Msida, Malta*
SRIRAM VIDYARTHI • *Department of Biological and Agricultural Engineering, University of California Davis, Davis, CA, USA*

Chapter 1

Methods and Protocols for Pulsed Electric Fields Treatment of Foods

Indrawati Oey, Stephen Giteru, and Sze Ying Leong

Abstract

The use of pulsed electric fields (PEF) technology to process foods has gained a broad range of industrial applications. PEF processing induces the permeabilization of the cell membrane and (micro)structural modification, which could enhance mass transfer process across the cell membranes leading to release of cell contents from the biological materials and microorganism inactivation. PEF-treated foods are considered superior in nutrient retention, color, and flavor. Furthermore, this technology uses low energy and has a low carbon footprint, which is favorable to the environment and sustainability. This article outlines important factors for handling the sample before and after PEF processing and how to safely operate PEF process. Various recommendations for assays to assess the treatment efficiency of PEF processing on solid foods are also discussed.

Key words Pulsed electric fields, Food, Electric field strength, Pulse frequency, Electroporation, Mass transfer, Non-thermal, Emerging food processing technology, Food quality, Texture

1 Introduction

Pulsed electric fields (PEF) processing applies high-voltage short-duration electric pulses (typically μs to ms) to food placed between two conducting electrodes. This processing induces the permeabilization of the cell membrane. When electric field is applied above a certain threshold intensity (i.e., the electric field strength is calculated based on the ratio between the peak electric potential difference (V) between two electrodes and the gap (cm) separating them [1, 2]), it leads to irreversible pore formation (usually known as "electroporation"). In general, increasing electric field strengths promotes the occurrence of pore formation. The occurrence of irreversible pore formation is mostly attained at electric field strengths above 0.5 kV/cm [3, 4], and it enhances the movement of ions and macromolecules, easing mass exchange between the medium and interior cells.

Mohsen Gavahian (ed.), *Emerging Food Processing Technologies*,
Methods and Protocols in Food Science, https://doi.org/10.1007/978-1-0716-2136-3_1,

PEF technology provides technological benefits not only as preservation (i.e., microorganism inactivation) but also as a processing aid to modify (micro)structure of food materials. The latter provides several industrial applications, for example, improving extraction yield (i.e., increasing the release of intracellular materials such as juice extraction, flavor, color, simple sugars, and bioactive compounds) [5], accelerating the maceration time in winemaking [6], enhancing nutritional value and digestibility [7], reducing cutting force [8], improving tissue flexibility [9], reducing oil absorption during frying [10], and increasing the rate of freezing, thawing, and (freeze) drying processes [11].

The efficiency and effectiveness of PEF processing are strongly dependent on the physicochemical properties of the food products used, sample preparation prior to PEF processing, PEF processing parameters, and the design of PEF equipment and chamber [12]. The electrical conductivity (or the resistance) of food products placed between the electrodes determines the effectiveness of transferring electric current through the food product. In addition, sample homogeneity influences the delivery of electric current, the efficiency of PEF processing, and the process uniformity [13]. Therefore, various precautions should be taken when unstable emulsion and the presence of solid particles or solid foods such as suspension, puree, mashed fruits, pulp in juice, fruit cuts in liquid, meat, whole fruits, and vegetables are used for PEF treatment.

This protocol outlines important factors related to sample handling before and after PEF processing and how to operate PEF processing safely. Various recommendations for assays to assess the treatment efficiency of PEF processing on solid foods are also discussed.

2 Procedures for PEF Processing

2.1 Sample Handling and Preparation Prior to PEF Treatment

The properties of samples greatly influence PEF processing and its effect [14]. Few important points must be taken into account, as listed below:

1. Assess the homogeneity and viscosity of the samples and measure the conductivity at the targeted initial temperature of PEF processing (*see* **Note 1**).
2. If the solid particles are incorporated in the sample mixture, check the stability of the mixture and avoid solid/liquid separation during the PEF processing (*see* **Note 2**).
3. When intact/whole solid foods are used, measure the electrical conductivity of the foods such as meat, potatoes, apples, etc., at the targeted initial temperature of PEF processing (*see* **Note 1**).

4. Be aware of food matrices with complex inherent fiber direction [8, 15]. When working with plant or animal muscle tissues, check the fiber direction (Figs. 1 and 2) with respect to the electric current direction (*see* **Note 3**).
5. An electrical current-conducting medium must fill any spaces between the intact/whole solid sample and the electrode for maximum sample-electrode contact. Measure the conductivity of the PEF conducting medium at the targeted initial temperature of PEF processing. Be aware that different solid-to-liquid ratio will affect the overall conductivity of the mixture (*see* **Note 4**).
6. Check the sample size and chamber dimension. Assess how to place the foods inside the chamber (either run as a batch system or as a continuous system) to enable maximum exposure to electric current. Food samples must be fully immersed in the PEF conducting medium. Record the sample position with respect to the electrode or electric current direction.
7. When foods are immersed in PEF conducting medium, ensure that the ionic strength of the buffer solution used does not result in passive diffusion.
8. Assess whether any handling and preparation steps applied to the sample before PEF processing will influence the stability or integrity of the analyte of interest or mask the targeted PEF

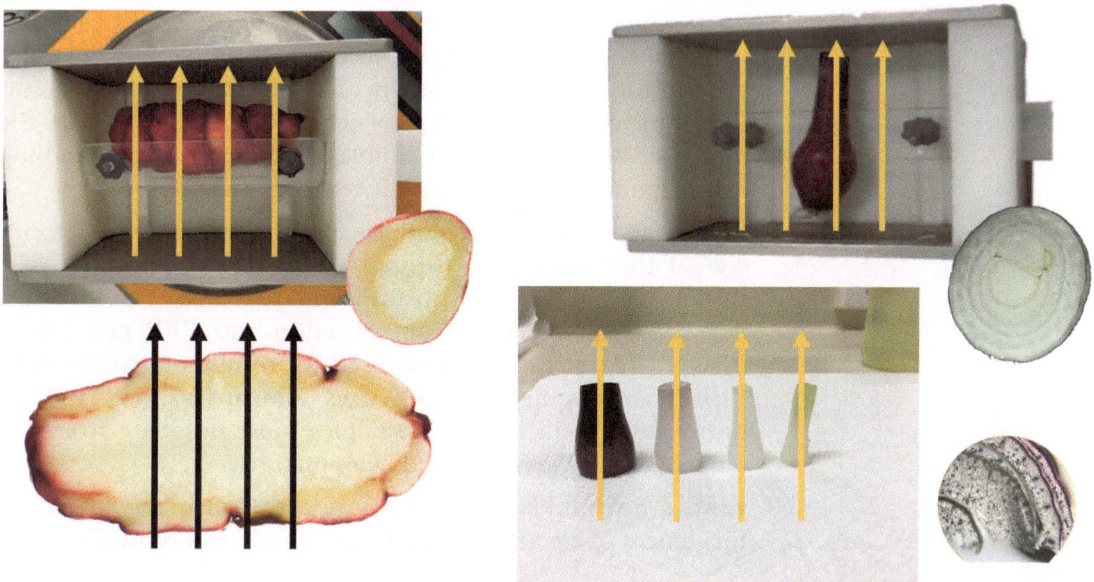

Fig. 1 Examples of positioning solid plant-based food tissues with a sample holder inside a PEF treatment chamber (left, oca tubers with different cell types from the outer region to the inner region; right, bunching onions made up of multiple similar organs), where the pulse current (in arrows) delivered across the samples. (The authors acknowledge Dr. Tingting Liu for providing the images (with permission))

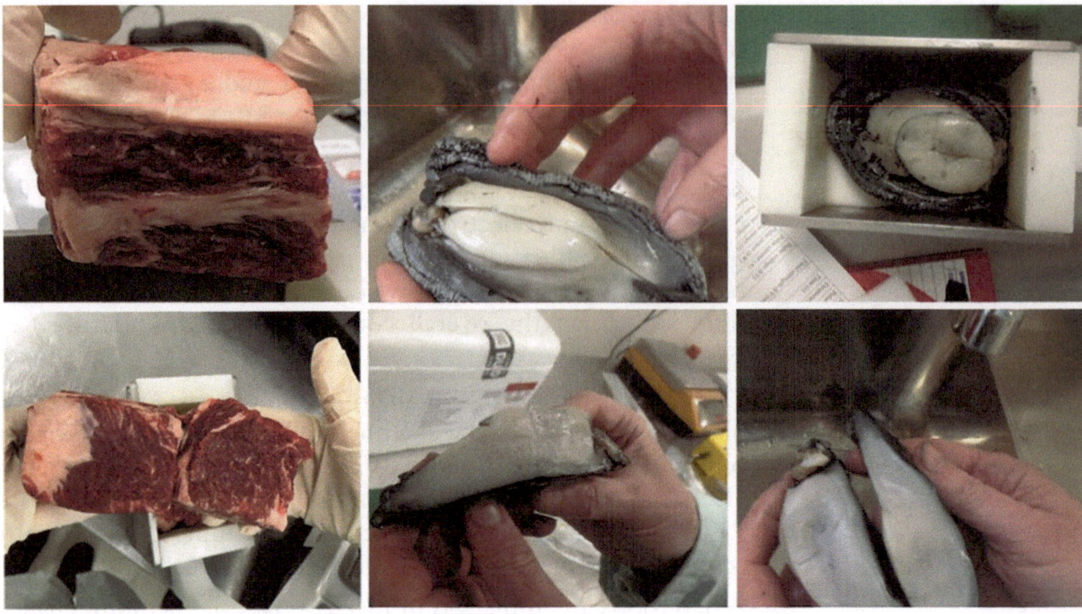

Fig. 2 Examples of solid muscle-based food tissues (left, beef muscle with heterogeneous distribution of connective tissues and fats; right, abalone with asymmetrical geometry) with inherent variation in cell types, structure, and muscle fibers. (The authors acknowledge Dr. Amali Alahakoon and Prof. Indrawati Oey for providing the beef muscle and abalone images, respectively (with permission))

effect. Be aware that peeling, cutting, slicing, mashing, or pureeing actions may induce mechanical damages, and initial leaching of cell contents may occur before applying PEF treatment on these sample types. Record the sample handling and preparation step in detail and include information on the methods or procedures applied prior to PEF treatment. Ensure the untreated and PEF-treated samples experience similar handling and preparation steps for result comparison.

9. Check the history of the samples used prior to PEF treatment, e.g., if they have been through the freezing/thawing cycle, storage condition, and postmortem time. Record the time between sample harvest or collection and PEF processing. Ensure the untreated sample used for result comparison with PEF-treated samples experiences similar history (or the untreated and PEF-treated samples must originate from the same batch of harvest, breed, age, maturity, etc.).
10. Ensure no bubble formation inside the chamber to reduce incidences of electrical breakdown of food (*see* **Note 5**).

2.2 Preparation of PEF Equipment

Before operating the equipment, it is important to conduct a thorough pre-start check on the safety features of the equipment, as outlined below. The equipment should not be connected to the mains electricity outlet during the safety check.

1. Mount the equipment safely by positioning it on a flat surface (larger equipment can be mounted on the floor).
2. Ensure that the space where the equipment is located is well ventilated and there is no obstruction around the equipment.
3. Ensure both the generator and treatment chamber (where the electrode is located) are not accessible during operation by installing suitable protective coverings (*see* **Note 6**).
4. Check the main power supply voltage and frequency to be connected to the equipment. Be aware of the location of electrical safety symbols and electrical hazard warning signs of the equipment that can cause serious personal injury or equipment damage if ignored.
5. Inspect that the equipment is designed with the electric current grounding feature.
6. Inspect the cooling medium level and the cooling function of the high-voltage transformer unit.
7. Check any water and oil leakage inside the equipment.
8. Inspect the electrodes of the treatment chamber for any wear out, changes in geometry, and/or an early sign of electrode corrosion. The external surface of the treatment chamber must be completely dry (*see* **Note 7**).
9. If using a continuous system, check the piping connections and conveyor belt system involved in channeling the sample through the PEF treatment chamber.

2.3 Operational Procedure for PEF Treatment Process

1. Establish good sample mixing and degassing conditions. Separation between solid and liquid should be avoided during PEF treatment (*see* **Note 2**).
2. (If appropriate) Pre-chill the sample to minimize the thermal effect induced during PEF when applied at high intensity. This step depends on the chemical composition and electrical properties of the sample and the experimental design of PEF treatment and the process intensity applied (*see* **Note 8**).
3. Load the sample into the treatment chamber (of known electrode gap/distance and geometry) and record the sample weight for specific energy calculation. The operator must ensure that air bubbles are not introduced during the filling process to minimize the incidences of dielectric breakdown. When solid foods are used, place the solid food in the chamber (of known electrode gap/distance and geometry) and fill the PEF treatment chamber with the conducting medium (e.g., buffer solution) at a specified solid-to-conducting medium ratio. It is recommended to centrally position the solid sample in the treatment chamber while matching the fiber orientation of foods (Figs. 1 and 2) relative to the direction of the electric current [8, 15] (*see* **Notes 2–4**).

4. Record the weight, initial electrical conductivity, and temperature of the sample for later monitoring of the PEF treatment process (*see* **Note 9**).
5. Configure the equipment with the input parameters required to achieve the desired effect (e.g., improve the efficacy of subsequent extraction, drying, freezing processes). Specify the pulse duration, pulse frequency, pulse number, and input pulse voltage and current of the generator (*see* **Note 10**).
6. Start the treatment process and record the attained electric field strength, pulse number, frequency, pulse width, pulse voltage, pulse current, pulse power, pulse energy, and resistance with the help of built-in sensors and measurement systems or external monitoring and measuring devices (*see* **Notes 11** and **12**).
7. Retrieve the sample from the chamber after the treatment and immediately measure the final electrical conductivity and temperature (*see* **Notes 13** and **14**).
8. Perform a visual examination of the sample immediately after the treatment to assess for undesired damage by electric fields, e.g., scorching of the food (cooked edges) and coagulation may indicate the occurrence of dielectric breakdown [16, 17] (*see* **Note 5**).
9. Post-PEF process handling steps depend on the specific application of PEF in food processing. The sample collection method should be standardized and appropriate to suit the end use or subsequent sample analysis.
10. Clean the treatment chamber with running tap water and mild washing detergent (avoid aggressive scrubbing and chemicals use), and then completely dry with a paper towel before subsequent use.

2.4 Operational Procedure for Continuous PEF Treatment Process

In contrast to the batch operation of PEF, a continuous flow process typically requires a larger volume of the sample before entering the treatment chamber to maintain a constant flow rate. Therefore, a suitable product delivery system is required to transport semisolid and solid samples. In the last 20 years, the continuous PEF process has been limited to liquid food products to achieve microbial and enzyme inactivation [18]. With the recent advancement in the design of pulse generators and treatment chambers, it is possible to treat an ample amount of semisolid (e.g., grape mash, olive mash) and solid (e.g., potato) food materials with PEF at a continuous flow. Due to the nature of the food material, semisolid foods (a mixture of solids and juice) are more easily transported in pipes (of suitable inner diameter) and then pumped into the treatment chamber [19–21]. In contrast, a conveyor belt system (insulator materials are highly recommended) is suitable for transporting solid foods into the PEF chamber [9].

2.4.1 Procedure

1. Before starting the machine, ensure all the auxiliary devices (e.g., peristaltic pump system, flow meter to monitor product flow rate, temperature probes/sensors are close to the electrodes) are connected correctly and a proper outlet that direct the treated sample into a sample collection tank is established.
2. Ensure the PEF machine is turned off.
3. Conduct regular cleaning-in-place (CIP) procedure according to [22] to clean and sanitize the entire product line from the feeding of product into the PEF treatment chamber and the sample collection outlet at the maximum possible pump speed.
   (a) Start with hot distilled water at ≥60 °C.
   (b) Follow by 1% (w/v) sodium hydroxide (CAS No.: 1310-73-2).
   (c) After that, pump in hot sterilized distilled water at ≥80 °C.
   (d) Follow by 1% (v/v) nitric acid (CAS No.: 7697-37-2).
   (e) Finally, pump in hot sterilized distilled water at ≥60 °C.
   (f) Each step is run for at least five times the volume of the entire product line.
   (g) *Optional:* Perform a surface swab of inlet and outlet of the product line for a quick microbiological test.
4. Mix the sample to attain a well-mixed sample and avoid air bubbles inside the PEF chamber. Semisolid foods (e.g., grape mash, olive mash with a mixture of solid and liquid components) must be well mixed at all times when pumping into the treatment chamber. If necessary, pre-chill the sample to minimize the thermal effect induced during PEF when applied at high intensity (*see* **Notes 2** and **8**).
5. Measure the electrical conductivity and temperature of the sample before the treatment (*see* **Note 9**).
6. Set up a constant volumetric flow in the treatment chamber (of known electrode gap/distance and geometry) where the sample fills up the entire product line with the help of the pump set at an appropriate flow rate (related to the residence time of the sample in the PEF treatment chamber) for sample delivery during PEF treatment. The operator must ensure that air bubbles are not introduced during the pumping process to minimize dielectric breakdown incidences and ensure the entire pipeline is free from an airlock (*see* **Note 5**).
7. Turn on the PEF machine and configure the equipment with operating parameters desired to achieve the treatment effect. Specify the pulse duration, pulse frequency, pulse number, and pulse voltage (*see* **Notes 10** and **15**).

8. Start the treatment process and record the attained electric field strength, flow rate, pulse number, frequency, pulse width, pulse voltage, pulse current, pulse power, pulse energy, resistance, and inlet and outlet temperatures with the help of built-in sensors and measurement systems or external monitoring and measuring devices (*see* **Notes 11**, **12**, **14**, and **16**).
9. Repeat the same for the untreated sample with the PEF generator turned off (*see* **Note 17**).
10. Collect the untreated and PEF-treated samples from the outlet and immediately measure the electrical conductivity and temperature (*see* **Notes 13** and **14**).
11. Standardize post-PEF process handling steps because leaching or microorganism (in)activation might continue during storage. Determining the time-lapse between PEF treatment and subsequent processing step such as extraction, separation, and frying is the most crucial step in post-PEF operations. This step depends on the targeted application of PEF in food processing.
12. At the end of the experiment, turn off the PEF generator and run the CIP procedure in the same sequence as outlined in **step 3**. If accessible, dry the treatment chamber entirely with a paper towel before subsequent use.

3 Procedures for Measuring the Process Parameters and Efficiency

3.1 Monitoring Electric Pulses Delivery During PEF Treatment

3.1.1 Procedure

1. Ensure all the required connections of the probe and digital storage oscilloscope (DSO) are fitted correctly to the PEF equipment.
2. Turn on PEF equipment and switch on DSO and the pulse signal recording/saving function.
3. Proceed with the PEF treatment of the sample at either batch or continuous configuration following the operational procedures described above.
4. Monitor the shape of pulses in the DSO during each treatment (*see* **Note 11**).
5. At the end of the experiment, determine the pulse parameters according to the pulse shape, as follows [23]:
   (a) *Exponential decay pulses*: described by maximum output amplitude value (voltage and current) and time constant (s) and the duration it takes for the pulse signal to drop from the maximum output value to 37% of the maximum value.
   (b) *Square wave pulses*: described by maximum output amplitude value (voltage and current) and the total time

(s) where the pulses reach 50% of the maximum output value at the signal rising and falling stages.

(c) *Other pulse shapes*: maximum output amplitude value (voltage and current), rise time (s), and fall time (s) where the time required for the pulse to increase from 10% to 90% of the output value and decrease from 90% to 10% of the output value, respectively.

6. Reconstruct/redraw the pulse signals delivered to the sample during each PEF treatment and present them as figures in the report. The figure can be either zoom in a single pulse or a segment of repetitive pulses delivered during the sample treatment or both.

3.2 Monitoring Output Voltage, Current, and Other Output Parameters During PEF Treatment

A suitable input and output monitoring system to monitor should be used to record and report PEF processing operating parameters during the entire course of the treatment. Some of the measured pulse parameters can be used to calculate the electric field strength (E, kV/cm) (Eq. (1)), energy density, or specific energy per pulse (Q, kJ/kg) (Eq. (2)) received by the treated product in the PEF chamber during each pulse [12], the electrical resistance of the sample in the treatment chamber (R, Ω) (Eq. (3)), pulse power (P_{pulse}, W) (Eq. (4)), pulse energy (E_{pulse}, J) (Eq. (5)), and treatment time (t, s) (Eq. (6)). Reporting Q value allows estimation of the energy consumption due to the PEF process and compares the treatment intensity when different types of PEF equipment are used. It is also worth noting that the (average) conductivity of the sample increases with electroporation and release of ions and macromolecules to the media (via leaching). Other factors such as the chamber geometry and configuration, pulse shape (square/exponential), treatment temperature, treatment time, and sample characteristics (i.e., molecular configuration and conductivity) play a crucial role in influencing the calculated output parameters of the PEF process [24, 25].

$$\text{Electric field strength, } E\ (\text{kV/cm}) = \frac{U}{\text{Electrode gap or distance}} \quad (1)$$

$$\text{Specific energy input per pulse}\left(Q, \frac{\text{kJ}}{\text{kg}}\right) = \frac{1}{m}\int_0^{\infty}\frac{U(t)^2}{R}\text{d}t = \frac{1}{m}\int_0^{\infty}U(t)\tilde{n}I(t)\text{d}t \quad (2)$$

$$\text{Electrical resistance, } R\ (\Omega) = \frac{\text{Electrode gap or distance}}{\text{Electrode area} \times \text{Electrical conductivity of sample}} \quad (3)$$

$$\text{Pulse power}, P_{\text{pulse}}\ (\text{W}) = I \times U \quad (4)$$

$$\text{Pulse energy for a square wave pulse}, E_{\text{pulse}}\ (\text{J}) = P_{\text{pulse}} \times \tau \quad (5)$$

$$\text{Treatment time}, t\ (\text{s}) = \tau \times n \quad (6)$$

where $U(t)$ and $I(t)$ are, respectively, the voltage (kV) across and current (A) through the treatment chamber load at time t, n is the number of pulses (dimensionless), m the sample weight (kg), τ is the pulse width (μs), and R is the electrical resistance (Ω).

Regarding the continuous flow of the PEF process, the specific energy input can be expressed according to Eq. (7). Estimating the residence time (t_r, s) (Eq. (8)) of the product in the treatment chamber during continuous PEF treatment of food materials and the total number of pulses (Eq. (9)) that the food material is exposed to, are critical process parameters.

$$\text{Specific energy input in a continuous system}, Q\left(\frac{\text{kJ}}{\text{kg}}\right) = \frac{f}{\dot{m}}\tilde{n}\int_0^{\tau} U(t)\tilde{n}I(t)\tilde{n}\text{d}t \quad (7)$$

$$\text{Product residence time}, t_r\ (\text{s}) = \frac{\text{Volume of treatment zone}}{\dot{m}} \quad (8)$$

$$\text{Pulse number in a continuous system}, \quad n = t_r\tilde{n}f \quad (9)$$

where f is the pulse frequency (Hz), $\dot{m}$ the mass flow rate of the product (kg/s), and t_r (s) is the product residence time.

3.3 Measurement of Temperature During or After PEF Treatment

An increase in sample temperature during PEF treatment is attributed to ohmic (Joule) heating. This phenomenon is significantly enhanced when high electric field strength, total specific energy, pulse frequency, and pulse width are applied to the sample to achieve higher electroporation of foods [12]. Fiber optic temperature sensors [26] are usually used to measure the temperature in the center of the sample in the PEF chamber. Multiple sensors need to be placed at both ends of each treatment chamber if several continuous flow treatment chambers connected in series are used. The locations of the temperature sensors in the PEF equipment should be clearly specified when reporting the result (*see* **Notes 14** and **16**).

4 Estimating the Degree of Cell Disintegration After PEF Treatment

It is crucial to determine whether the applied PEF process parameters are adequate to cause cell electroporation or pore formation to the biological tissue. As a result of PEF treatment, the food sample is expected to experience subtle damage/disintegration, allowing cell contents to continuously pass through the electroporated cell membrane into the intercellular spaces, thereby changing the electrical and structural properties of the sample. Estimating the cell disintegration degree of PEF-treated food sample will aid the selection of optimal PEF process parameters to achieve specific desired outcomes. Several approaches can be used to estimate the cell disintegration degree of food sample after being subjected to a PEF treatment process: direct measurement of electrical conductivity of sample before and after PEF treatment [27], determination of the cell disintegration index to estimate the proportion of permeabilized cells [28], determination of electrolyte/ion leakage to measure intactness and permeability of the cell membranes [29], monitoring the leakage of specific compounds from the treated sample as a function of time after PEF treatment [30], and structural analysis.

4.1 Direct Measurement of Electrical Conductivity Before and After PEF Treatment

4.1.1 Procedure

1. Calibrate the conductivity meter against a standard calibration solution(s) with a known conductivity value.
2. Rinse the electrode of the conductivity meter with distilled water.
3. Measure and record the conductivity and temperature of the sample before PEF treatment at multiple locations of the food sample.
4. Rinse the electrode of the conductivity meter.
5. Proceed with sample treatment in the PEF equipment following the operational procedures described above.
6. Immediately after PEF, measure and record the conductivity and temperature of the sample at multiple locations of the food sample.

   If the sample is suspended in a conducting medium, the electrical conductivity of the medium should be measured before and immediately after PEF and the evolution followed as a function of time after PEF to be related with the leakage of cell contents.
7. Record all the measurements and express the result as Siemens per centimeter (S/cm). Report the electrical conductivity value for at least four independent food samples per PEF process treatment.

4.2 Estimation of the Cell Disintegration Index Through Electrical Conductivity Measurement (Suitable for Solid Foods)

4.2.1 Procedure

1. Repeat the steps above to obtain the electrical conductivity measurement at multiple sample locations before and after PEF treatment.
2. To prepare a completely disintegrated sample, immediately freeze the PEF-treated sample (−20 °C for 24 h or longer depending on the food materials used) followed by thawing at room temperature (~23 °C for 12 h).
3. Measure and record the conductivity and temperature of the completely disintegrated samples.

4. Calculate the electrical conductivity disintegration index (Z_c, dimensionless) according to Eq. (10):

$$Z_c = \frac{\sigma - \sigma_i}{\sigma_f - \sigma_i} \tag{10}$$

where σ, σ_i, and σ_f (S/cm) are the electrical conductivity of PEF-treated sample, sample before PEF treatment, and completely disintegrated sample, respectively. The above equation gives $Z_c = 0$ for the untreated sample and $Z_c = 1$ for the utterly disintegrated sample [28].

5. Report the measured values of σ_i and σ_f and the calculated Z_c value averaged from at least four independent food samples per PEF process treatment.

4.3 Estimation of the Cell Disintegration Index Through Impedance Measurement (Suitable for Solid Foods)

This is a method based on the frequency dependence of electrical conductivity of intact and permeabilized biological tissue.

4.3.1 Procedure

1. Using an impedance analyzer, calibrate the instrument according to the manufacturer's instruction (*see* **Note 18**).
2. Configure the range of low- and high-frequency fields for sample measurement and ensure the instrument is equipped with a data acquisition function to allow the subsequent time and frequency domain analysis.
3. Proceed with the sample treatment using PEF as described above.
4. For solid PEF-treated samples, after PEF treatment, cut the sample into a uniform dimension of either cylinder (e.g., 1 × 1 cm, diameter × length) or disc form. Sampling should be done at multiple locations of the solid sample. For a semi-solid sample, obtain a fixed volume for the impedance measurement [31].
5. Perform the impedance measurement of the untreated and PEF-treated samples at different parameters (e.g., frequency levels in the range of 1 kHz to 50 MHz).
6. Calculate the cell disintegration index based on impedance measurement (Z_p, dimensionless) according to Eq. (11):

$$Z_p = 1 - \left(\frac{C_h}{P_h}\right) \times \frac{P_h - P_l}{C_h - C_l} \tag{11}$$

where C_h and C_l (S/cm) are the electrical conductivities of the sample before PEF treatment measured in high-

(3–50 MHz) and low-frequency fields (1–5 kHz), respectively. P_h and P_l (S/cm) are the electrical conductivities of the sample after PEF treatment measured in high- (3–50 MHz) and low-frequency fields (1–5 kHz), respectively. The above equation gives $Z_p = 0$ for the untreated sample and $Z_p = 1$ for the utterly disintegrated sample [32]. A completely disintegrated sample (e.g., expose a sample to at least two freeze/thaw cycle) of similar dimension/volume can be prepared as a reference to ensure $Z_p = 1$.

7. Report the measured values of C_h, C_l, P_h, and P_l, along with the calculated Z_p value averaged from at least four independent food samples for each PEF treatment.

4.4 Determination of Electrolyte Leakage (Suitable for Solid Plant Foods Only)

One of the critical steps in this assay involves identifying a suitable isotonic solution (e.g., mannitol) and concentration to equilibrate the sample immediately after PEF treatment [33]. After that, the electrical conductivity of the solution incubated with the PEF-treated sample is measured as a function of time. This assay allows the collected data to be interpreted in two ways: (1) the percentage of electrolyte leakage when the electrical conductivity of the isotonic solution incubated with the PEF-treated sample reaches a plateau and (2) fitting of the electrolyte leakage time data into a suitable model and the relevant kinetic parameters will be estimated [34]. However, this assay is not suitable for plant tissues with thick cell wall and low electrolyte content.

4.4.1 Procedure

1. Prepare mannitol (CAS No.: 69-65-8) solutions at different molarities, e.g., ranging from 0 to 0.8 M.
2. Use a cork borer to obtain representative tissue samples of uniform dimension and weight from multiple locations of the food.
3. Weigh the samples and transfer each of them into individual 50 mL polypropylene tubes.
4. Fill each tube with the mannitol solution at different molarities at a solid-to-liquid ratio of approximately 1:10 (w/v). Soak the sample in each mannitol solution for up to 5 h at room temperature.
5. After 5 h of soaking, remove the samples from the mannitol solutions, blot them dry using a paper towel, and then weigh and record the weight.
6. Repeat the above steps using at least five samples for each mannitol concentration.
7. Choose the mannitol concentration that does not result in a weight change of the sample. This is the most suitable isotonic solution for equilibrating the sample after PEF treatment.

8. Proceed with sample treatment using PEF as described above. Immediately after PEF treatment, remove the sample from the treatment chamber, rinse with distilled water for 30 s to remove the surface electrolytes, blot dry using paper towels, and weigh the sample (cut the sample into uniform dimension if necessary). Repeat the same for the untreated sample.
9. Transfer the untreated and PEF-treated samples into individual polypropylene containers prefilled with the isotonic mannitol solution (as determined in **steps 1–7**).
10. Ensure that the sample is fully immersed in the isotonic solution at a ratio of 1:10 (w/v) and incubate at either room temperature, controlled-temperature incubator (25 °C), or constant temperature water bath (25 °C).
11. Measure the electrical conductivity of the isotonic mannitol solution periodically for up to 5 h. It is recommended to establish a measurement interval of 5 min in the first hour and then every 30 min for the remaining 4 h. Gentle agitation of the solution with the sample is recommended between measurements.
12. At the end of the 5 h incubation, freeze the entire polypropylene container (sample and mannitol solution) overnight at −20 °C and then thaw the sample the next day at room temperature. Repeat the freeze–thaw procedure at least two times. Finally, measure the electrical conductivity of the fully thawed mixture to obtain the total conductivity.
13. Calculate the percentage of electrolyte leakage using Eq. (12).

$$\text{Electrolyte leakage }(\%) = \frac{\text{Electrical conductivity at incubation time } t}{\text{Total conductivity}} \times 100\% \tag{12}$$

14. Report the percentage of electrolyte leakage when the electrical conductivity of the isotonic solution incubated with the PEF-treated sample reaches a plateau (e.g., usually t between 240 min and 300 min). Report the percentage of electrolyte leakage for at least four independent food samples per PEF treatment process. Alternatively, calculate the percentage of electrolyte leakage throughout the 5 h incubation time and fit the data into a growth saturation model based on the Michaelis-Menten equation (Eq. (13)) [34, 35] to estimate the maximum percentage of electrolyte leakage and the time required to reach 50% of the maximum percentage electrolyte leakage.

$$y = \frac{a \times t}{b + t} \tag{13}$$

where y is the percentage of electrolyte leakage calculated from Eq. (12) at specific incubation time t, a is the asymptote that represents the maximum percentage of electrolyte leakage reached by the sample, and b is the time to reach 50% of the maximum percentage of electrolyte leakage. A sample with a low estimated b value suggests the maximum percentage of electrolyte leakage is achieved more rapidly.

4.5 Monitoring the Leakage of Cell Contents of a Sample as a Function of Time After PEF Treatment (Suitable for Solid Foods Only)

When estimating the degree of cell disintegration in a sample caused by PEF, the most common approach is to conduct immediate measurement on the PEF-treated sample, as described in Subheading 4.1. However, irreversible pore formation induced by PEF will allow free movement of cell contents and various organelles within the cell into the intercellular space and outside of the cell to the surrounding. A natural mass transfer process would occur, and thus the leakage of a large number of cell contents is expected to occur either instantaneously or gradually (within hours or days) depending on the raw material matrix and the applied intensity of PEF treatment.

4.5.1 Procedure

1. Observe the leakage of the inherent liquid component from the sample or a reduction of specific major compounds of the sample within a fixed time point after PEF treatment.

   For example, comparing the final yield and the chemical composition of juice collected from untreated and PEF-treated grapes [36], observing the reduction of oxalate antinutrient from PEF-treated oca tubers [33], and determining the amount of purge (water) loss and tissue shrinkage for animal muscle tissue after PEF treatment by calculating the weight loss over the initial sample weight [15].
2. Place the PEF-treated food material into a suitable isotonic solution (e.g., mannitol at the most optimum concentration) and determine cell leakage by measuring the concentration of specific compounds inherent to the food leached out into the isotonic solution at specified time intervals.
3. Using the observed leakage of cell contents, develop kinetics information on the rate of leakage and the maximum leakage duration as a function of time.

4.6 Microstructural Analysis (Suitable for Solid Foods)

4.6.1 Procedure

1. Prepare representative specimens (either whole mounts, fixated, embedded, sectioned, or stained) from untreated and PEF-treated food material to examine for any alterations in the structure at a microscopic level under a microscope.
2. Use microscopic imaging techniques such as light, fluorescent, electron (SEM, cryo-SEM, TEM), and confocal laser scanning microscopic analysis or optical coherence tomography technique to identify the structural changes in PEF-treated specimen.

3. Follow the guidelines below in performing microstructural analysis and reporting on PEF-treated samples.
   (a) The dimension of the sample preparation for the microscopic analysis.
   (b) The time delay between the completion of PEF treatment to specimen preparation should be kept minimal and included in the report.
   (c) The methodology for each specimen preparation step: mounting, fixation, embedding, sectioning, and staining.
   (d) The rationale of the choice of staining dye (if applicable).
   (e) Information of the microscope (model name, manufacturer, magnification, and settings) and the camera (model name, manufacturer, and camera settings) to capture the micrographs.
   (f) The number of sample replication.

4.7 Texture Analysis (Suitable for Solid Foods Only)

An application of PEF treatment may induce changes in the textural properties of biological tissue, and thus comparing the texture parameters for untreated and PEF-treated foods should be considered.

4.7.1 Procedure

1. Use a texture analyzer with an appropriate penetration/cutting probe suitable for the food sample to conduct the test.
2. Adopt a generic texture profile analysis (TPA—based on double compression test) capable of quantifying multiple textural parameters (i.e., hardness, cohesiveness, springiness, and resilience) in one texture measurement. Such information will allow the researchers to compare the difference in textural parameters between untreated and PEF-treated samples.
3. The following guidelines provide information that should be obtained when performing the texture analysis and included in reporting the method and results.
   (a) The dimension of the sample preparation for the texture analysis.
   (b) The time delay between the completion of PEF treatment to texture measurement should be kept minimal and included in the report.
   (c) The rationale of the choice of cutting probe (include the model name, size, diameter).
   (d) Information of the texture analyzer: model, manufacturer, calibration, weight of load cell, dimensions of the probe, compression and withdrawal speed, compression strain, compression distance, and positioning of the sample on the platform.

(e) Choose the most appropriate textural parameters to describe the food sample.

(f) The number of sample replication.

4.8 Determination of the Treatment Homogeneity by Considering the Biochemical Changes of Food Sample After PEF Treatment

Since plant tissues are composed of different cell types and sizes, it becomes critical to determine whether all cell types are uniformly exposed to PEF. Cell electroporation triggered by PEF treatment could influence the movement of compounds between cellular compartments, causing some changes in biochemical reactions within the plant tissue. Two simple biochemical-related assays proposed to determine the homogeneity of PEF treatment of plant tissues include (1) viability of cells using the staining method and (2) the interaction between endogenous enzymes and substrates due to decompartmentalization of organelles as the cell ruptured after PEF. With respect to animal muscle tissues, a considerable deviation in pH values and increased conductivity of the PEF-treated sample compared to the untreated sample is a quick indication of significant structural and biochemical changes in the muscle tissue [15].

4.8.1 Cell Viability Staining Technique Using Neutral Red (NR) (Based on Passive Diffusion of the Dye Across the Tonoplast Membrane of the Vacuole of Plant Tissue, Intact Plant Cells Appear as Dark Red Color)

Procedure

1. Prepare the neutral red (NR) stock dye solution by dissolving 0.5% NR (3-amino-7-dimethylamino-2-methylphenazine hydrochloride, CAS No.: 553-24-2) in acetone for 30 min and then filter twice using Whatman No. 1 filter paper.
2. On the day of the experiment, dilute the NR dye in a suitable buffer (preferably in an isotonic solution) to an appropriate concentration (e.g., dilute the stock solution to 0.04% using isotonic solution of 0.2 M mannitol or 0.01 M HEPES (*N*-(2-hydroxyethyl)piperazine-*N*′-2-ethane sulfonic acid) buffer of pH 7.8) [29].
3. Proceed sample treatment with PEF as described above.
4. After PEF, immediately cut a specimen and rinse it in deionized water to remove cell debris due to the cutting action.
5. Dip the specimen in diluted NR dye solution for at least 2 h.
6. Rinse the stained specimen in the isotonic solution (the same buffer used to dilute the NR dye in **step 2**) for at least 30 min.
7. Mount the specimen on a microscope slide and add a drop of deionized water.
8. Observe the specimen with a light microscope and capture the micrographs with a digital color camera. Repeat the same steps for the untreated sample.
9. Process and analyze the micrographs with image analysis software to calculate the proportion of red-colored area (i.e., indicate intact cells) over the total uncolored (i.e., indicate cells lose integrity and viability) and colored areas.

10. Report the proportion area of cells with intact integrity and viability by averaging from at least four independent food samples per PEF process treatment.

4.8.2 Cell Viability Staining Technique Using Tetrazolium Salt (Based on Reduction of Oxidoreductase Enzymes in the Mitochondria to Form Red Formazan in Plant Tissue, Intact Plant Cells Appear as Dark Red Color)

Procedure

1. On the day of the experiment, prepare 0.5% w/v tetrazolium (2,3,5-triphenyltetrazolium chloride, CAS No.: 298-96-4) salt solution in isotonic solution.
2. Proceed sample treatment with PEF as described above.
3. After PEF, immediately obtain a specimen and rinse in deionized water to remove cell debris due to the cutting action.
4. Transfer the specimen into a petri dish prefilled with tetrazolium solution.
5. Ensure that the specimen is fully immersed in the tetrazolium solution at a ratio of 1:5 (v/v), cover the petri dish with a lid, and incubate in a dark controlled temperature incubator (20 °C) for at least 30 min. Repeat the same steps for the untreated sample (*see* **Note 19**).
6. After staining, rinse the stained sample section with deionized water and blot dry with a paper towel.
7. Set up a copy stand unit and LED copy lights on both sides and place the digital camera at a fixed position above the base of the copy stand unit.
8. Place the stained sample section in the middle and capture the image using a digital camera. Appropriate background color should be used to show the contrast of the stain and tissue color.
9. Process and analyze the images with image analysis software to calculate the proportion of red-colored area (i.e., indicate intact viable cells) over the total uncolored (i.e., indicate cells lose viability) and red-colored areas [30].
10. Report the proportion area of cells with intact integrity and viability by averaging from at least four independent food samples per PEF treatment process.

4.8.3 Visual Examination of the Enzymatic Browning Area (Based on Enzymatic Browning Reaction Between Endogenous Polyphenol Oxidase and Phenol Compounds in Plant Tissue, Permeabilized Plant Cells Appear as Dark Brown)

Procedure

1. Proceed sample treatment with PEF as described above.
2. After PEF, immediately cut a specimen from the sample and rinse it in deionized water to remove cell debris due to the cutting action.
3. Transfer the specimen into a petri dish and add a few drops of deionized water to the sample section to prevent dehydration.
4. Leave the petri dish uncovered and exposed to the surrounding to facilitate the enzymatic browning reaction for at least a few hours. Repeat the same steps for untreated sample (maintain same incubation time for untreated and PEF-treated sample).
5. Set up a copy stand unit and LED copy lights on both sides and place the digital camera at a fixed position above the base of the copy stand unit.
6. Place the sample section in the middle and capture the image using a digital camera.
7. Process and analyze the images with image analysis software to calculate the proportion of dark brown-colored area (i.e., indicate permeabilized cells) over the total uncolored (i.e., indicate intact cells) and dark brown-colored areas [13].
8. Report the proportion area of permeabilized cells by averaging at least four independent food samples per PEF process treatment.

5 Monitoring the Condition of Electrodes in the PEF Treatment Chamber

The treatment chamber is a crucial component of the PEF equipment, where the samples receive high-voltage electric pulses after coming into contact with the electrodes. However, the electrode condition may deteriorate with time and extended usage. The following procedures can be used to monitor the overall electrode condition and to detect electrode corrosion.

5.1 Procedure Involving Routine Inspection of the Electrode

1. Perform a visual check on the appearance of the electrode.
2. Measure the thickness of the electrodes, using a gauge with high precision, before and after each experiment.
3. Weigh the treatment chamber before and after each experiment.
4. The electrode should be replaced immediately if any signs of damage or corrosion are observed at this stage.

5.2 Procedure Involving Quantification of Metal Ions

1. For every experiment, prepare three specimens: (1) PEF-treated, (2) untreated (i.e., a sample from the same preparation step, brought to contact with the PEF electrode but without PEF exposure, where the pulse generator is turned

off), and (3) control food samples (i.e., from the same preparation step and never in contact with the PEF electrode) (*see* **Note 20**).

2. Completely digest all food samples using concentrated nitric acid and heating (90–95 °C) [37]. If the PEF treatment (usually batch configuration) was conducted with a food sample suspended in a conducting medium (e.g., buffer), the medium should be collected before and after PEF treatment to quantify the concentration of metal ions migrated from the electrode to the buffer.
3. For the analysis of metal ions: calibrate the equipment of choice with certified standards (e.g., inductively coupled plasma mass spectrometry (ICP-MS) or atomic absorption spectrophotometry (AAS)). Specify the limit of detection (LOD) for each analyzed metal ion during reporting, as outlined in AOAC Official Method 2015.01.
4. Determine the concentrations of metal ions of interest, e.g., iron (Fe), chromium (Cr), nickel (Ni), manganese (Mn), platinum (Pt), and titanium (Ti), usually depending on the electrode material in contact with the food sample.
5. Determine the difference between the concentration of metal ions measured in the PEF-treated and untreated samples, representing the actual metal concentration released from the electrodes to the food sample.
6. Compare the concentration of each metal element detected in the food sample against the recommended upper limit of daily dietary intakes for adult consumers and the local legislation values of metal concentration in food for human consumption to ensure that the levels of metal contaminants do not cause toxicity.

5.3 Procedure Involving Microscopic Inspection of the Electrode Surface

1. Using a scanning electron microscope with adequate magnification and equipped with a camera-capturing function, obtain the micrographs of the following surfaces:
   (a) Ground electrode before PEF processing.
   (b) Ground electrode after PEF processing.
   (c) High electric field applying electrodes after PEF processing.
2. Compare the changes in surface morphologies, increase in surface roughness and porosity, and the presence of globular bumps between the electrodes [38].

6 Notes

1. The electrical conductivity of a food material is highly dependent on the temperature. It is recommended to calibrate the conductivity meter at the same temperature as the sample to be measured or use a temperature-compensated conductivity meter. Record the start and final conductivity (S/cm) and the starting/final temperature of the sample (°C).
2. Mixing followed by degassing of viscous foods using de-aeration units helps to eliminate air cells and therefore enable uniform distribution of electric fields that helps to minimize dielectric breakdown [1, 16].
3. Plant-based food samples have variation in their chemical composition, cell types, and structure (Fig. 1). The presence of fibers interferes with the delivery of electric fields and temperature development leading to the development of blisters. On the other hand, muscle-based food (e.g., beef and fish) presents a variation in chemical composition (fat and protein content) and structure (connective tissue content) (Fig. 2) [39].
4. PEF treatment of solid foods requires a conductive treatment medium such as water, buffer, or a gel with conductivity and/or ionic strength closely matched to that of the sample. Filling the chamber with a suitable conductive medium increases the treatment efficiency by allowing maximum contact between the electrodes and the sample. It also improves the transmission of electrical current, reduces the risk of localized heating and development of air bubbles that could lead to arching, regulates the sample conductivity, regulates the starting temperature, and controls the ionic strength of the food matrix or adjustment of sample viscosity in semisolid foods [40]. Prepare an adequate volume of conductive medium with uniform electrical conductivity and pH to maintain uniformity between treatments. Typical conductive media, their conductivity, and food characteristics in various conductive media are summarized in Table 1.
5. *Dielectric breakdown*: It is usually characterized by a spark and occurs when the applied electric field strength exceeds the dielectric strength of the treated food in the chamber. This phenomenon is also referred to as arcing [42]. Dielectric breakdowns can also be initiated by the enhancement of local electric fields by impurities (highly conducting particles) or air spaces (gas bubbles) due to differences in dielectric properties [1, 43, 44]. Disadvantages of dielectric breakdown include non-uniform electric field distribution, scorching, ionization, generation of shock waves, and damaging electrodes.

Table 1
Conductivity of solid food samples in treatment media

Material	Medium concentration	Conductivity (S/m) at 25 °C
Sodium chloride (g/L)	0.02 0.05 0.17	0.19 0.55 1.79
Sodium phosphate (g/L)	0.03 0.05 0.10	0.19 0.36 0.67
Potato	Raw	0.32
	Water	0.25
	0.2% NaCl	0.37
	0.4% NaCl	0.36
	0.8% NaCl	0.43
Carrot	Raw	0.13
	Water	0.12
	0.2% NaCl	0.29
	0.4% NaCl	0.31
	0.8% NaCl	0.25
Yam	Raw	0.11
	Water	0.09
	0.2% NaCl	0.42
	0.4% NaCl	0.35
	0.8% NaCl	0.35
Chicken	Raw	0.37
Beef	Raw	0.44

Adapted from [41]

Recognition of the occurrence of dielectric breakdown during PEF treatment: It is characterized by the following events: a bright luminous spark, formation of pits on the electrodes, evolution of bubbles, and explosive sound due to the release of pressure from a liquid. Other ways in which the electrical breakdown of foods can be identified is through an increase in the flashover count if a flashover detection unit is installed in the equipment [17].

Avoiding dielectric breakdown during PEF treatment: Application of electric fields in the treatment chamber containing bubbles results in the expansion and spherical elongation of the gas (more than five times) contained in the bubble. This amplifies electric fields at the ends of the sphere, exceeding the dielectric strength of the gas bubble [42]. The localized peaks of electric fields volatilize the surrounding liquid, which produces vapor that may cause the bubble to expand further and discharge. The growth of the bubble can bridge the gap between the two electrodes producing a spark. The presence of bubbles in semisolid foods (high viscosity) can be minimized by homogenization and degassing of the material before processing [16]. A simple de-aeration unit to accomplish this task may consist of two stainless steel containers, pressure gauges, and a vacuum pump [45]. In general, a guiding principle to prevent overcoming dielectric breakdown thresholds at a given pressure is to target bubbles of <0.5 mm when using a chamber of 3 mm gap [42]. For optimum results, the setup process must address the required homogenization and vacuum-degassing conditions to assure uniformity of the material and the absence of trapped air that promote dielectric breakdown [46]. Other ways of reducing the incidence of dielectric breakdown include the use of smooth electrode surface, PEF treating food at a lower temperature, cooling the electrodes, use of round-edged electrodes, and ensuring uniform distribution of electric fields within the food by proper chamber design [1, 47]. The PEF process operating parameters must be set below the dielectric breakdown threshold [48].

Corrective action in case of a dielectric breakdown during PEF treatment: When an electrical breakdown occurs, stop the PEF treatment process and turn off the equipment. Clean any spillage and inspect the treatment chamber and electrode for potential damage.

6. A protective covering acting as a Faraday shield to contain the electromagnetic fields within the treatment chamber, thus protecting the user against exposure to the high-intensity electric fields, is required.
7. Generally, uniform electric fields are delivered using parallel plate electrodes with a gap sufficiently smaller than that of the electrode surface area. Sharp-edged electrodes should be avoided due to the potential accumulation of surface charge density in the sharp edges, which can cause dielectric breakdown of food [43, 49]. Disc-shaped, round-edged electrodes are preferable to minimize the possibility of a dielectric breakdown during PEF [50, 51].
8. The chemical composition of a sample is directly related to its electrical properties, and thus a high conductive region within the semisolid and solid sample is likely to promote an increase in local temperature or free radical generation [52]. On the contrary, when a low-intensity PEF treatment is applied (e.g.,

aiming at enhanced mass transfer), the sample might experience a sudden temperature increase and then a quick drop to equilibrate with the surrounding medium in the treatment chamber during the gap between pulses and after PEF treatment due to energy dissipation. A cooling system for the electrodes (e.g., recirculating a temperature-regulated coolant) can minimize temperature increase during PEF treatment [49, 53]. Otherwise, it is generally recommended to pre-cool the samples to 4 °C before PEF to minimize the possible thermal damage that can occur to the food during the treatment. Moreover, placing the PEF equipment in a ventilated and cold room can allow rapid dissipation of the heat converted from electrical energy during PEF.

9. The sample composition, conductivity, ionic strength, pH of the treatment medium, water activity, moisture content, and sample weight within and between batches must be standardized to maintain reproducibility. Record the start and final conductivity (S/cm) and the starting/final temperature of the sample (°C).

10. The intensity of PEF treatment delivered across the sample placed between the electrodes is regulated by certain input parameters, including the pulse shape (square or exponential), polarities (bipolar, monopolar), pulse width (duration of single pulse), pulse energy, pulse repeat rate (frequency), and number of pulses (determines the total time of sample exposure). The harmonized quantities of the delivered PEF intensity are the electric field strength and specific energy [12]. These depend on the voltage applied, treatment time, and resistance of the treatment chamber that varies according to the geometry and conductivity of the treated material.

11. It is highly recommended to compare the input and output pulse signals by installing the digital storage oscilloscope (DSO) probes as close as possible to the electrodes, with no additional connecting wires. The frequency bandwidth of the probe (within the range of MHz or at least five times the maximum signal bandwidth) should match with the DSO to allow accurate measurement and improve the resolution of the micro- or millisecond pulse signals typical for most PEF applications. Another important feature of the DSO is the rise time specification to capture accurate details of the short yet fast pulse signals. Moreover, the sampling rate of a DSO should be adequate to allow real-time pulse display and monitoring during PEF treatment in order to observe the overall pulse shapes (i.e., exponential decay or square wave pulses) and whether the pulses have deviated from the expected forms. A DSO capable of recording and saving information about the pulse signals for subsequent analysis or signal reconstruction is advantageous.

12. For custom-built PEF equipment without an integrated measurement system, it is necessary to configure an external device (e.g., Rogowski coil to measure high-speed current pulses, flow meter to measure volumetric flow rate of liquid sample when entering the treatment chamber) and use an oscilloscope with compatible high-voltage and current probes (positioned as close as possible to the electrodes/treatment chamber) to measure the relevant pulse parameters. It is recommended to conduct periodic calibration, as part of the equipment maintenance, on the built-in and customized sensor and measurement systems to obtain robust pulse parameters after each PEF treatment.
13. The electrical conductivity of a food sample can increase linearly with an increase in temperature during ohmic heating. Furthermore, the conductivity of a sample increases with increasing input voltage, which is explained by the electroporation phenomenon and the release of solutes/cell contents.
14. An accurate temperature measurement of the sample during pulses delivery (in micro- and millisecond range) is challenging and almost impossible based on the currently available temperature monitoring technology. Direct temperature measurements using conventional thermocouples placed inside the food sample when high-intensity electric fields are applied in the treatment chamber will cause interference to the PEF treatment. The sample temperature is usually measured as the average temperature after equilibration with the surrounding environment. Apart from a fiber optic temperature sensor, the use of an infrared thermal sensor array to record the real-time temperature distribution of sample inside the chamber during PEF treatment could be a practical alternative.
15. If the experiment conducted in a continuous PEF treatment configuration involves testing several treatment intensities on the same batch of the sample, perform the highest electric field strength treatments first, followed by the lower electric fields. This could avoid running the CIP procedure between each PEF treatment of different intensities. This step is crucial when PEF is used for microorganism inactivation.
16. Very often, continuous PEF application for microbial and enzyme inactivation requires the food sample to receive a preheating treatment in a heat exchanger before entering the PEF treatment chamber. Thus, the temperature of a sample should be measured before preheating, after preheating, and before entering the treatment chamber. Moreover, PEF-treated food samples are immediately cooled when exiting the PEF chamber, and thus the temperature of the sample after PEF and

before and after the heat exchanger should be measured and reported.

17. To ensure a fair comparison of the effect of PEF (conducted at continuous configuration) on sample characteristics, any mechanical damages that occurred to the sample due to the pumping system and delivery into the restricted diameter of the pipeline should be taken into consideration.
18. An inductance (L), capacitance (C), and resistance (R) analyzer (LCR) meter can be used as an alternative to measure the electrical characteristic of a sample before and after PEF treatment [54].
19. Tetrazolium solution is very sensitive towards light, pH change, and susceptible to precipitation. Maintain the same incubation time for untreated and PEF-treated sample. Overnight sample incubation in the tetrazolium solution is needed for some plant tissue due to the difference in cell wall structure.
20. Previous studies have demonstrated that deterioration of electrode can be accelerated by repetitive application of high-intensity electric field strength, long pulse width, and high pulse frequency on high conductive food sample [55–57]. Under high-intensity electric field strength, the electrode material is oxidized and corroded due to electrochemical reactions. Electrode corrosion will trigger the release/migration of metal ions into the food sample, triggering food safety problems. There is generally no consensus on using a specific type of electrode material across all the PEF equipment model. However, titanium, platinized titanium, and stainless steel are commonly reported in most studies for both laboratory and commercial-scale PEF equipment due to their excellent corrosion resistance [38].

Acknowledgments

Oey and Leong are affiliated to the Riddet Institute, a New Zealand Centre of Research Excellence, funded by the Tertiary Education Commission. This work is supported by the Food Industry Enabling Technologies (FIET) program funded by the New Zealand Ministry of Business, Innovation and Employment (contract MAUX1402).

References

1. Barbosa-Cánovas GV, Góngora-Nieto MM, Pothakamury UR, Swanson BG (1999) Chapter 1—Fundamentals of high-intensity pulsed electric fields (PEF). In: Barbosa-Cánovas GV, Góngora-Nieto MM, Pothakamury UR, Swanson BG (eds) Preservation of foods with pulsed electric fields. Academic Press, San Diego, pp 1–19

2. Zhang Q, Barbosa-Cánovas GV, Swanson BG (1995) Engineering aspects of pulsed electric field pasteurization. J Food Eng 25(2):261–281
3. Angersbach A, Heinz V, Knorr D (2000) Effects of pulsed electric fields on cell membranes in real food systems. Innov Food Sci Emerg Technol 1(2):135–149
4. Bazhal MI, Lebovka NI, Vorobiev E (2003) Optimisation of pulsed electric field strength for electroplasmolysis of vegetable tissues. Biosyst Eng 86(3):339–345
5. Donsì F, Ferrari G, Pataro G (2010) Applications of pulsed electric field treatments for the enhancement of mass transfer from vegetable tissue. Food Eng Rev 2(2):109–130
6. Leong SY, Burritt DJ, Oey I (2016) Effect of combining pulsed electric fields with maceration time on merlot grapes in protecting Caco-2 cells from oxidative stress. Food Bioprocess Technol 9(1):147–160
7. Abduh S, Leong SY, Agyei D, Oey I (2019) Understanding the properties of starch in potatoes (*Solanum tuberosum* var. Agria) after being treated with pulsed electric field processing. Foods 8(5):159
8. Leong SY, Richter LK, Knorr D, Oey I (2014) Feasibility of using pulsed electric field processing to inactivate enzymes and reduce the cutting force of carrot (*Daucus carota* var. Nantes). Innov Food Sci Emerg Technol 26: 159–167
9. Fauster T, Schlossnikl D, Rath F, Ostermeier R, Teufel F, Toepfl S, Jaeger H (2018) Impact of pulsed electric field (PEF) pretreatment on process performance of industrial French fries production. J Food Eng 235: 16–22
10. Liu T, Dodds E, Leong SY, Eyres GT, Burritt DJ, Oey I (2017) Effect of pulsed electric fields on the structure and frying quality of "kumara" sweet potato tubers. Innov Food Sci Emerg Technol 39:197–208
11. Barba FJ, Parniakov O, Pereira SA, Wiktor A, Grimi N, Boussetta N, Saraiva JA, Raso J, Martin-Belloso O, Witrowa-Rajchert D (2015) Current applications and new opportunities for the use of pulsed electric fields in food science and industry. Food Res Int 77: 773–798
12. Raso J, Frey W, Ferrari G, Pataro G, Knorr D, Teissie J, Miklavčič D (2016) Recommendations guidelines on the key information to be reported in studies of application of PEF technology in food and biotechnological processes. Innov Food Sci Emerg Technol 37:312–321
13. Oey I, Faridnia F, Leong SY, Burritt DJ, Liu T (2016) Determination of pulsed electric fields effects on the structure of potato tubers. In: Miklavcic D (ed) Handbook of electroporation. Springer International Publishing, Cham, pp 1–19
14. Barsotti L, Merle P, Cheftel J (1999) Food processing by pulsed electric fields. I. Physical aspects. Food Rev Int 15(2):163–180
15. Faridnia F, Bekhit AEDA, Niven B, Oey I (2014) Impact of pulsed electric fields and post-mortem vacuum ageing on beef *longissimus thoracis* muscles. Int J Food Sci Technol 49(11):2339–2347
16. Chauhan OP, Unni LE, Toepfl S (2014) Quality of high-pressure processed pastes and purees. In: Tokusoglu Ö, Swanson BG (eds) Improving food quality with novel food processing technologies. CRC Press, Boca Raton, pp 111–131
17. Alahakoon AU, Faridnia F, Bremer PJ, Silcock P, Oey I (2017) Pulsed electric fields effects on meat tissue quality and functionality. In: Miklavcic D (ed) Handbook of electroporation. Springer International Publishing, Cham, pp 2455–2475
18. Vega-Mercado H, Martín-Belloso O, Qin B-L, Chang FJ, Marcela Góngora-Nieto M, Barbosa-Cánovas GV, Swanson BG (1997) Non-thermal food preservation: pulsed electric fields. Trends Food Sci Technol 8(5):151–157
19. Leong SY, Treadwell M, Liu T, Hochberg M, Sack M, Mueller G, Sigler J, Silcock P, Oey I (2020) Influence of pulsed electric fields processing at high-intensity electric field strength on the relationship between anthocyanins composition and colour intensity of merlot (*Vitis vinifera* L.) musts during cold maceration. Innov Food Sci Emerg Technol 59:102243
20. Puértolas E, López N, Saldaña G, Álvarez I, Raso J (2010) Evaluation of phenolic extraction during fermentation of red grapes treated by a continuous pulsed electric fields process at pilot-plant scale. J Food Eng 98(1):120–125
21. Goettel M, Eing C, Gusbeth C, Straessner R, Frey W (2013) Pulsed electric field assisted extraction of intracellular valuables from microalgae. Algal Res 2(4):401–408
22. Sharma P, Bremer P, Oey I, Everett DW (2014) Bacterial inactivation in whole milk using pulsed electric field processing. Int Dairy J 35(1):49–56
23. Pirc E, Reberšek M, Miklavčič D (2017) Dosimetry in electroporation-based technologies and treatments. In: Markov M (ed) Dosimetry in bioelectromagnetics. CRC Press, Boca Raton, pp 233–268

24. Jaeger H, Meneses N, Knorr D (2014) Food technologies: pulsed electric field technology. In: Motarjemi Y (ed) Encyclopedia of food safety, vol 3. Academic Press, Waltham, pp 239–244
25. Toepfl S, Heinz V, Knorr D (2007) High intensity pulsed electric fields applied for food preservation. Chem Eng Process 46(6):537–546
26. Espachs-Barroso A, Van Loey A, Hendrickx M, Martín-Belloso O (2006) Inactivation of plant pectin methylesterase by thermal or high intensity pulsed electric field treatments. Innov Food Sci Emerg Technol 7(1–2):40–48
27. Leong SY, Oey I (2014) Effect of pulsed electric field treatment on enzyme kinetics and thermostability of endogenous ascorbic acid oxidase in carrots (*Daucus carota* cv. Nantes). Food Chem 146:538–547
28. Lebovka NI, Bazhal MI, Vorobiev E (2002) Estimation of characteristic damage time of food materials in pulsed-electric fields. J Food Eng 54(4):337–346
29. Ersus S, Barrett DM (2010) Determination of membrane integrity in onion tissues treated by pulsed electric fields: use of microscopic images and ion leakage measurements. Innov Food Sci Emerg Technol 11(4):598–603
30. Faridnia F, Burritt DJ, Bremer PJ, Oey I (2015) Innovative approach to determine the effect of pulsed electric fields on the microstructure of whole potato tubers: use of cell viability, microscopic images and ionic leakage measurements. Food Res Int 77(3):556–564
31. Angersbach A, Heinz V, Knorr D (1999) Electrophysiological model of intact and processed plant tissues: cell disintegration criteria. Biotechnol Prog 15(4):753–762
32. Janositz A, Knorr D (2010) Microscopic visualization of pulsed electric field induced changes on plant cellular level. Innov Food Sci Emerg Technol 11(4):592–597
33. Liu T, Burritt DJ, Eyres GT, Oey I (2018) Pulsed electric field processing reduces the oxalate content of oca (*Oxalis tuberosa*) tubers while retaining starch grains and the general structural integrity of tubers. Food Chem 245:890–898
34. Asavasanti S, Ersus S, Ristenpart W, Stroeve P, Barrett DM (2010) Critical electric field strengths of onion tissues treated by pulsed electric fields. J Food Sci 75(7):E433–E443
35. Liu T, Burritt DJ, Oey I (2018) Understanding the effect of pulsed electric fields on multilayered solid plant foods: bunching onions (*Allium fistulosum*) as a model system. Food Res Int 120:560–567
36. Leong SY, Burritt DJ, Oey I (2016) Evaluation of the anthocyanin release and health-promoting properties of pinot noir grape juices after pulsed electric fields. Food Chem 196(1):833–841
37. Demirel S, Tuzen M, Saracoglu S, Soylak M (2008) Evaluation of various digestion procedures for trace element contents of some food materials. J Hazard Mater 152(3):1020–1026
38. Kim MJ, Zhang HQ (2010) Improving electrode durability of PEF chamber by selecting suitable material. In: Zhang HQ, Barbosa-Cánovas GV, Balasubramaniam VMB, Dunne CP, Farkas DF, Yuan JTC (eds) Nonthermal processing technologies for food. Wiley-Blackwell: IFT Press, Chichester, pp 201–212
39. Töpfl S (2006) Pulsed electric fields (PEF) for permeabilization of cell membranes in food- and bioprocessing—applications, Process and equipment design and cost analysis. Doctoral thesis, Technische Universität Berlin
40. Alirezalu K, Munekata PES, Parniakov O, Barba FJ, Witt J, Toepfl S, Wiktor A, Lorenzo JM (2020) Pulsed electric field and mild heating for milk processing: a review on recent advances. J Sci Food Agric 100(1):16–24
41. Palaniappan S, Sastry SK (1991) Electrical conductivities of selected solid foods during ohmic heating. J Food Process Eng 14(3):221–236
42. Góngora-Nieto MM, Pedrow PD, Swanson BG, Barbosa-Cánovas GV (2003) Impact of air bubbles in a dielectric liquid when subjected to high field strengths. Innov Food Sci Emerg Technol 4(1):57–67
43. Arshad RN, Abdul-Malek Z, Munir A, Buntat Z, Ahmad MH, Jusoh YMM, Bekhit AE-D, Roobab U, Manzoor MF, Aadil RM (2020) Electrical systems for pulsed electric field applications in the food industry: an engineering perspective. Trends Food Sci Technol 104:1–13
44. Zimmermann U (1986) Electrical breakdown, electropermeabilization and electrofusion. In: Pedersen SHF (ed) Reviews of physiology, biochemistry and pharmacology, vol 105. Springer Berlin Heidelberg, Berlin, Heidelberg, pp 175–256
45. Alkhafaji SR, Farid M (2007) An investigation on pulsed electric fields technology using new treatment chamber design. Innov Food Sci Emerg Technol 8(2):205–212
46. Toepfl S, Siemer C, Heinz V (2014) Chapter 8 - effect of high-intensity electric field pulses on solid foods. In: Sun DW (ed) Emerging technologies for food processing, 2nd edn. Academic Press, London, pp 147–154

47. Monfort S, Mañas P, Condón S, Raso J, Álvarez I (2012) Physicochemical and functional properties of liquid whole egg treated by the application of pulsed electric fields followed by heat in the presence of triethyl citrate. Food Res Int 48(2):484–490
48. Haberl-Meglic S, Levicnik E, Luengo E, Raso J, Miklavcic D (2016) The effect of temperature and bacterial growth phase on protein extraction by means of electroporation. Bioelectrochemistry 112:77–82
49. Lindgren M, Aronsson K, Galt S, Ohlsson T (2002) Simulation of the temperature increase in pulsed electric field (PEF) continuous flow treatment chambers. Innov Food Sci Emerg Technol 3(3):233–245
50. Ho S, Mittal GS (2000) High voltage pulsed electrical field for liquid food pasteurization. Food Rev Int 16(4):395–434
51. Buckow R, Ng S, Toepfl S (2013) Pulsed electric field processing of orange juice: a review on microbial, enzymatic, nutritional, and sensory quality and stability. Compr Rev Food Sci Food Safe 12(5):455–467
52. Raso J, Palop A, Pagán R, Condón S (1998) Inactivation of *Bacillus subtilis* spores by combining ultrasonic waves under pressure and mild heat treatment. J Appl Microbiol 85(5):849–854
53. Luengo E, Martinez JM, Coustets M, Alvarez I, Teissie J, Rols MP, Raso J (2015) A comparative study on the effects of millisecond- and microsecond-pulsed electric field treatments on the permeabilization and extraction of pigments from *Chlorella vulgaris*. J Membr Biol 248(5):883–891
54. Asavasanti S, Ristenpart W, Stroeve P, Barrett DM (2011) Permeabilization of plant tissues by monopolar pulsed electric fields: effect of frequency. J Food Sci 76(1):E98–E111
55. Gad A, Jayaram SH (2014) Effect of electric pulse parameters on releasing metallic particles from stainless steel electrodes during PEF processing of milk. IEEE Trans Ind Appl 50(2):1402–1409
56. Pataro G, Barca GMJ, Pereira RN, Vicente AA, Teixeira JA, Ferrari G (2014) Quantification of metal release from stainless steel electrodes during conventional and pulsed ohmic heating. Innov Food Sci Emerg Technol 21:66–73
57. Roodenburg B, Morren J, Haan SWHD, Berg HE (2003) Corrosion experiments in a pulsed electric field (PEF) treatment chamber with stainless steel electrodes fed with three different pulse shapes. In: Pulsed power seminar proceedings, Loughborough, UK, 23 September 2003, p 7

Chapter 2

Cold Plasma Processing: Methods and Applications in Study of Food Decontamination

Manreet Bhullar, Mohsen Gavahian, and N. N. Misra

Abstract

Cold plasma is a partially ionized gas including ions, electrons, radicals, photons, and neutral molecules. The exploration of cold plasma as an emerging food processing technology continues to increase in research laboratories, particularly for sterilization application. This chapter outlines protocols for handling of food samples for cold plasma sterilization and electrical and optical characterization of the plasma. As a use case, all methods are outlined for a dielectric barrier discharge (DBD) plasma setup.

Key words Cold plasma, Food decontamination, Disinfection, Electrical discharge

1 Introduction

The ionized state of a gas, plasma, is an excellent antimicrobial agent and has scientifically proven efficacy against bacteria, yeasts, molds, and viruses. Plasma generation can be carried out under atmospheric pressure conditions and at ambient temperatures in a variety of form factors for treatment of solid as well as liquid substrates—foods, biological suspensions/solutions, water, and even gases [1]. Of the many uses of cold plasma treatments under exploration for the food industry, the decontamination application is of high interest to the food researchers. Widespread exploration of cold plasma technology with acceptable microbial load reduction, insignificant quality changes, and good retention of shelf life have been observed for fresh-cut fruits and vegetables like apples, melons, blueberry, strawberry, pears, cucumber, tomatoes, lettuce, etc. [2, 3]. Promising results of cold plasma have also been observed in decontamination of several other food products, including meat and meat products [4–6], dairy [7], cereals [8, 9], and fruit juices [10, 11]. These positive microbicidal effects of cold plasma are due to the cocktail of reactive chemical species generated through ionization of the gas [12]. However, cold plasma

Mohsen Gavahian (ed.), *Emerging Food Processing Technologies*,
Methods and Protocols in Food Science, https://doi.org/10.1007/978-1-0716-2136-3_2,

treatment is not applicable to all products, especially those rich in sensitive lipids [13].

The efficacy of plasma assisted decontamination is dictated by several factors, some of which are peculiar to each type of the plasma equipment. For the present discussion, it is assumed that the readers are familiar with the basics of plasma technology and plasma sources (discussed elsewhere; e.g., Misra et al. [14]). While there are many different types of plasma sources, including corona, dielectric barrier discharge (DBD), microwave plasma, gliding arc, etc., the DBD is most employed in food research. A popular configuration of the DBD also included the in-package plasma technology, where the packaged food is subjected to cold plasma treatment [3]. Therefore, we will confine the scope of this chapter to protocols for use of DBD plasma in decontamination of foods.

2 Materials

2.1 Preparation of Food Samples and Microbial Cultures

When working with fresh produce or short shelf life food samples, the samples are recommended to be not stored for more than 24 h prior to experimentation. The time lag between harvesting and experimentation should be based on the judgment of physiological changes in the produce. Any necessary storage must only be carried out under refrigerated conditions and with appropriate packaging. The products, if required, must be cut into appropriate equivalent sized pieces prior to plasma treatments, e.g., 10–12 mm diameter disks of carrot [2]. The carrots must be grouped according to experimental design (e.g., control and treatment classes).

The preparations required for studying the antimicrobial effects of cold plasma would depend on the objective of the study. In general, following are the two common cases:

Case 1: Studying the antimicrobial efficacy of cold plasma against natural microflora of the food.

Case 2: Studying the antimicrobial efficacy of cold plasma against bacteria artificially inoculated on to the food surface.

The steps in preparation of the food material for each case will be described hereunder, taking the example of a fresh produce, say tomatoes, lettuce, or strawberries.

Case 1: Efficacy against native microflora

1. The first step for quantifying the effect of any antimicrobial treatment against the natural microflora of the food is to determine the control levels of the microflora. Several number of test foods/items are tested for evaluating the number of microflora population present on the food.

2. The treated test foods can be enumerated for presence of different microflora using different media types and incubating plates at specific temperature and humidity conditions (*see* Subheading 3.3 post treatment microbial enumeration).

Case 2: Efficacy against inoculated microflora

1. As a first step, it is essential that the absence of the targeted bacteria (say, *E. coli* or *Listeria* spp.) in the food sample is confirmed.
2. A spot inoculation is generally recommended for solid foods. Spot inoculation can be carried out with 0.1 mL of diluted bacterial inoculum to result in ca. 6 $\log_{10}$ CFU/g population, as a general recommendation. However, there are several studies that have conducted research using dip inoculation methods as well.
3. The inoculated samples must be allowed to dry in a sterile environment (e.g., a laminar flow hood) or under refrigeration temperature for at least 1 h to ensure cell attachment [7, 15].

2.2 Preparation of the Plasma Source

A DBD plasma system typically comprises of a pair of metallic (aluminum or stainless steel) electrodes of well-defined geometry (circular or rectangular shaped). The electrodes are covered with a dielectric material to limit the charge. Example of the dielectric material includes polypropylene or quartz. A high voltage in the order of several kilovolts is applied across the metallic electrodes to induce the ionization of the gas trapped within the inter-barrier volume. The power supply to a DBD is invariably an alternating current (AC) source or a pulsed direct current (DC) source and could oscillate at frequencies ranging from 50 or 60 Hz to kilo or mega-Hz [2, 3]. The samples for treatment could either be placed directly on the dielectric material or inside a package within the inter-electrode volume. In Fig. 1, example of a model DBD plasma setup for in-package plasma treatment of foods is included along with dimensions and materials of construction.

Prior to execution of the plasma treatment studies, it is to be ensured that the ground connections of the DBD are intact and the voltage regulator is set to zero. If the samples are to come in direct contact with the dielectric material, these should be sterilized using UV treatment or other methods (*see* **Note 1**). The entire plasma system must ideally be placed inside a laboratory laminar flow chamber.

If ambient air is employed as the working gas for plasma discharges, the humidity and temperature of the ambient environment must be recoded using a RH/temperature meter and used for reporting purposes. This is important considering that the humidity and gas temperature dictate the plasma chemistry.

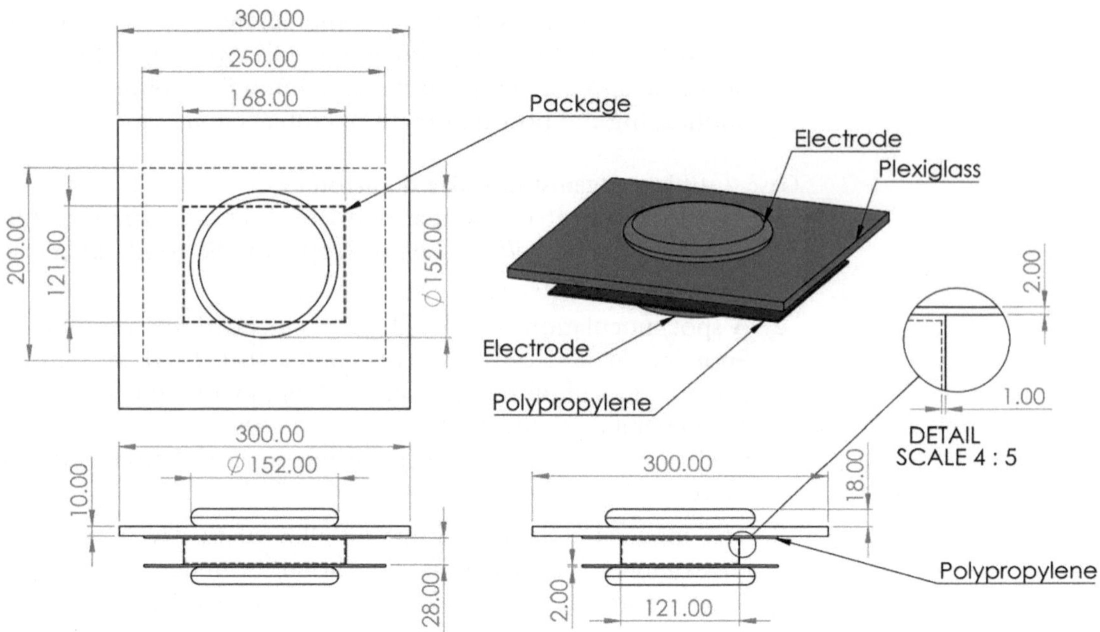

Fig. 1 Model DBD plasma setup for in-package plasma treatment of foods (isometric view on top-right). The package and dielectrics are made of polypropylene in this exemplary figure. The dielectric barriers could alternatively be made of quartz or thin ceramic sheet

2.3 Preparation of Packages and/or Sample Holders

1. Always place the samples intended for plasma treatment in pre-sterilized containers or sterile disposable containers. The sample holders must be made of non-conductive materials, such as glass or food-grade plastic, as they will be exposed to high-voltage environment.
2. If in-package plasma treatment is to be carried out, the packages must be sterilized under UV light for 1–4 h and the samples loaded inside a laminar flow cabinet. Care must be taken to avoid the use of any metallic pins, considering that sharp points could act as corona initiators in presence of strong electric fields.
3. The use of alcohol or other chemical methods should not be practiced for sterilization of the sample holders or boxes, as the chemical vapors could alter the plasma chemistry and, thus, the experimental results.

3 Methods

3.1 Inoculation of Samples

1. Fresh produce should be purchased immediately before the treatment and should not be stored more than 48 h at refrigerated temperature before the experimentation (*see* **Note 2**).
2. Diverse microorganisms are naturally present on fresh produce which attributes to the natural microflora. For the studies

Fig. 2 (**a**) Dip inoculation of strawberries; (**b**) spot inoculation of tomato

focusing on inactivation of specific human pathogens on produce, rinsing of test fruits and vegetables with water to remove any soil debris followed by treatment with 70% ethanol [16], rinsing with 100 ppm free chlorine solution for 5 min [17, 18], and dipping in hot water for 10 s are recommended to inactivate any background flora.

3. The antimicrobial-resistant strain of target organism of interest can also be used to negate the effect of background flora in the cell enumeration process. Cell cultures of target pathogen should be subcultured twice in appropriate conditions of growth media, incubation temperature, and time as per the Bacteriological Analytical Manual (BAM) methods. After centrifugation of the growth culture and double washing of the cells pellets, harvested cells can be resuspended in phosphate buffer saline (PBS) to make a working inoculum of 8–9 $\log_{10}$ CFU/mL, which should be confirmed with plating method.
4. The produce can be inoculated in two ways: spot inoculation (a smaller surface area is inoculated) (e.g., 2 cm^2) with 50–100 μL of inoculum volume or dip inoculation (the whole produce is dipped in appropriate inoculum solution for few minutes to attach the cells) (*see* Fig. 2). The samples can be air-dried after this (*see* **Note 3**).

Dip Inoculation Procedure: Dip Inoculation of Fresh Produce (*see* **Note 4**)

1. Propagate a 48-h subculture of the desired inoculation organism following BAM protocol.
2. After incubation, wash the cells with 1× phosphate buffer saline (PBS) and adjust to a required final volume of the stock inoculum.

3. In a sterile Whirl-Pak bag, prepare the inoculum at a sufficient volume and concentration to completely submerge the produce.
4. Flame sterilize tongs, transfer the produce to the Whirl-Pak bag containing the inoculum, and completely submerge the produce for at least 10 min.
5. After 10 min, flame sterilize the tongs, remove the produce from the inoculum, and place the produce onto a sterile petri dish or a tray for a 30–60-min drying period at ambient temperature in the biosafety cabinet.
6. After drying, flame sterilize the tongs and rotate the produce onto a dry sterile petri dish for the inoculum on the bottom of the produce to dry. The dried produce is now ready for experimentation.

Spot Inoculation Procedure: Similar to dip inoculation method, the spot inoculation is done by depositing low volumes (10–100 μL) onto a known surface area of the produce. Spot inoculation is usually done by forming small droplets using a pipette and letting it dry under the biosafety cabinet for 30–60 min, unlike dip inoculation where the whole produce is submerged in the inoculum.

3.2 Plasma Treatment of Food Samples

1. Place the inoculated samples in the sterilized plasma equipment, if subjecting to open air plasma treatment. Alternatively, for in-package plasma treatment, package the samples inside sterilized food-grade rigid or flexible packages. Subsequently, place the packaged samples in the dielectric barrier discharge plasma setup (*see* **Note 5**).
2. Turn on and operate the plasma equipment as per the guidelines of the equipment supplier.
3. Perform the plasma treatment as per the experimental design, involving varying of the process parameters such as voltage (or electric field), treatment time, and power.
4. For microbial count estimations, the plasma treated and control samples can be stored under refrigerated conditions for 24 h prior to analysis. If subjecting to in-package plasma treatment, the packaged samples must be stored for an appropriate duration under refrigerated or room temperature conditions. For example, a storage for 24 h under refrigerated conditions was suggested for cut carrots subjected to in-package plasma treatment by Mahnot et al. [2].
5. At the end of treatment period, the microbial load enumeration must be carried out for control and plasma-treated samples to assess the efficacy of each treatment condition.

Fig. 3 Bacterial enumeration method: the treated produce is dipped in BPW solution to harvest the survivor cells from the produce surface

3.3 Enumeration of Treated Food Samples

1. The treated sample can be enumerated in different ways for determining the survivor populations of the inoculum, either using log reduction per unit surface area (cm^2) or per fruit (grams).
2. Treated produce is transferred to the 0.1% (w/v) buffered peptone water in volume relevant to the size of the test produce such that the produce submerges in the BPW solution in the Whirl-Pak (stomacher) bag (Fig. 3).
3. Whirl-Pak (stomacher bag) containing the treated and untreated produce can be hand massaged for 30 s to 1 min, pulsified using a pulsifier to remove the attached cells, or stomached (if the fruits size is small or the produce is cut into pieces).
4. The processed sample bag can be diluted tenfold making several dilution tubes, and 100 μL from each tube is plated onto agar media in duplicate.
5. For aerobic plate counts, Tryptic soy agar with yeast extract media (TSAYE) can be used. For yeast and mold count, use Dicholoran Rose Bengal Chloramphenicol agar (DRBC agar) media. In addition, nutrient agar incubated at 30 °C for 24–48 h in air can be used for aerobic bacteria and in modified atmospheres (5 kPa O_2 and 20 kPa CO_2) for facultative aerobic bacteria; Potato Dextrose Agar (PDA) with the addition of chloramphenicol (100 μg/mL) incubated at 30 °C for 48 h for yeast and fungi; Lactobacilli MRS broth with addition of Bacto™ agar (10 g/L), incubated at 30 °C for 72 h under modified atmosphere (20 kPa CO_2 and 5 kPa O_2) for lactic acid bacteria; and MacConkey agar incubated at 37 °C for 24 h for enteric bacteria. A CO_2 water-jacketed incubator with

automatic gas atmosphere composition control can be used for creating the modified atmosphere (*see* **Notes 6** and 7).

6. A colony counter or manual counting of the plates can be done for enumeration of colonies.

3.4 Electrical Characterization of the Plasma Discharge

1. During experiments, monitor the voltage applied to the electrodes using a suitable voltage divider or high-voltage probe connected to an oscilloscope. A recommended voltage divider could include a 1000:1 resistor bank (*see* **Note 8**).
2. If required, monitor the discharge current using a suitable current transformer (CT) connected to a high-resolution oscilloscope. The ground wire should pass through the center of the current transformer. The turns ratio of the CT may require some preliminary trials for initial setup, especially when the current is very low (in the order of micro-amperes). If the intention is to capture the current pulse events, a Rogowski coil can be used. Rogowski coils are typically toroidal in geometry, where the alternating magnetic field produced by the current carried in wire induces a voltage in the coil that is proportional to the rate of change. If one wishes to wind a Rogowski coil, care must be taken that the winding must be as even as possible to avoid susceptibility to magnetic pickup in adjacent conductors or other sources of magnetic fields.
3. Monitor the power consumed and the current drawn by the entire plasma system using a wall power meter. If required the effectiveness of the power system can be assessed using the power factor value, typically displayed on wall power meters.
4. In order to obtain an estimate of the actual power fed to the gas in discharge gap, we recommend the capacitive voltage division method, originally developed by Manley (Manley, 1943). The details of the method are as follows:
   (a) Connect a capacitor in series with the ground electrode and the ground. The value of this measurement capacitor, $C_m \gg$ CDBD, the capacitance of the DBD. Often a high-voltage capacitor in the nF range would be sufficient for most DBDs.
   (b) Using an appropriate voltage divider or high-voltage probe, record the voltage waveform across the measurement capacitor (C_m) using an oscilloscope. A 100:1 probe would be sufficient for this purpose (*see* **Note 9**).
   (c) As a next step, plot the voltage applied to the electrodes against the voltage across external capacitor. This will yield a closed form Lissajous figure that is ideally elliptical to parallelogram in shape.

(d) Calculate the area of the Lissajous figure after appropriate smoothing of the signals, if needed. The area can alternatively be calculated using Gauss's method. The total power can be calculated using the following formula:

$$P = 2 * \text{pi} * f_s * C_m * A_{liss}$$

where P is the power in W, pi is the constant equivalent to 3.14, f_s is the switching frequency (generally 50 or 60 Hz), C_m is the measurement capacitance in F, and A_{liss} is the area of the Lissajous figure.

(e) A MATLAB code is provided in Box 1 as a guidance for the calculations. Herein the approach involves carrying out Savitzky-Golay filtering to smooth out the voltage signals, followed by identifying the zero crossing points. Subsequently, one complete cycle of waveform is taken for calculation of the area via Gauss's method.

Box 1 MATLAB Script to Calculate the Power Fed to Plasma Discharge Using Lissajous Method

```
% MATLAB script for calculating power fed to the discharge
volume by Lissajous method
% Author: NN Misra
% Date: April 18, 2021
% Ver: 3.2

% The script takes data from an excel file where first
column contains a common time vector
% second column contains the applied voltage (Vapp)
% third column contains voltage across the measurement
capacitor (Vc = 6 nF)
% subsequent columns alternate with Vapp and Vc for
various experimental conditions

clear; close all; clc; % start with a clean workspace
p = xlsread('power_data.xlsx'); % read the excel file
t = p(:,1); % extract time column

%% iterate over the pair of columns
for ii = 2:2:size(p,2)

 x = p(:,ii); % extract applied voltage data, Vapp
 y = p(:,ii+1); % extract voltage across capacitor, Vc
  y = sgolayfilt(y,1,51); % 51 length savitzk-golay
filtering of current data
```

(continued)

```
% find zero-crossings - use the crossings function
available from file exchange
% Steffen Brueckner (2020). crossing
% https://www.mathworks.com/matlabcentral/fileexchange/
2432-crossing
% MATLAB Central File Exchange. Retrieved May 12, 2020.
[indV,t0V] = crossing(x,t);
[indC,t0C] = crossing(y,t);

% Truncate between first and third zero crossing
(corresponding to one cycle)
trunT_voltage = t(indV(1):indV(3));
trun_voltage = x(indV(1):indV(3));

trunT_current = t(ismember(t,trunT_voltage));
trun_current = y(ismember(t,trunT_voltage));

npoints = length(trunT_voltage) - length(trunT_current);

% make vectors of equal length, if that is not the case
if npoints >= 0
trunT_voltage = trunT_voltage(1:end-npoints);
trun_voltage = trun_voltage(1:end-npoints);
else
trunT_current = trunT_current(1:end+npoints);
trun_current = trun_current(1:end+npoints);
end

x = trun_voltage; % Vapp ready for calculation
y = trun_current; % Vc ready for calculation

%%% area calculation by gauss method
a_gauss = 0; % initialize gaussian area variable
va = x;
vC = y;
for i = 1:length(x)
x1 = va(i);
y1 = vC(i);
if (i<length(x))
x2 = va(i+1);
y2 = vC(i+1);
end
a_gauss = a_gauss + x1*y2 - x2*y1; % calculate area
end
a_gauss = a_gauss/2.0; % end of gauss method
```

(continued)

```
Cm = 6e-9; % capacitance of measurement capacitor in F
freq = 60; % Frequency of applied voltage
power = a_gauss*Cm*2*pi*freq % calculate power
end
```

3.5 Study of Plasma Chemistry

To characterize the reactive oxygen and nitrogen species (RONS) generated from the plasma source optical emission spectroscopy (OES) is widely employed. To obtain the emission spectrum, of the plasma discharge, following protocol is to be implemented in a dark room (*see* Fig. 4 for reference):

1. Place an optical lens at a suitable horizontal distance of 5–15 cm away from the plasma discharge, such that the focal point of the lens lies within the discharge volume. If the lens holder is of metallic construction, care should be taken to avoid it from acting as a ground against the high voltage.
2. Place the end of an optical fiber at the other end (focal point) of the lens. For weak emissions, a 1000 μm core fiber optic cable is recommended. Ensure that the optical fiber is suitable for UV-VIS spectroscopy and is solarized (to handle UV light). The optical fiber should not be bent more than the recommended bending radius.
3. Connect the optical fiber to the light acquisition port of a calibrated spectrophotometer. Examples of spectrophotometers commonly used for plasma spectroscopy include the HR2000/HR2000+/HR4000 from Ocean Insight (formerly, Ocean Optics), Black-Comet spectrometers from StellarNet, AvaSpec-Mini spectrometers from Avantes, etc. [19].

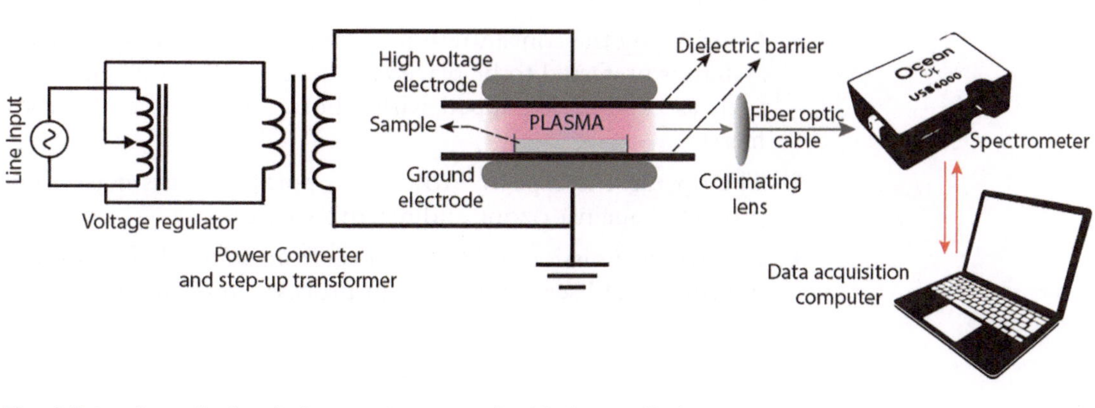

Fig. 4 Setup for optical emission spectroscopy of cold plasma discharges

4. In the software for spectrometer control, input appropriate values of integration time, averaging, and the total spectrum acquisition time. Recommended values that have been found to give satisfactory results for high voltage plasma discharges include an integration time of 5 s and averaging of 5–10 spectra.
5. Considering that there is always dark noise that overlaps a signal in a spectrometer, a correction has to be implemented. For this, collect spectrum in dark room without exposure to any light. The dark noise should not follow any pattern or exhibit any characteristic peaks.
6. Subsequently, acquire spectrum of the plasma discharge under the experimental conditions. Subtract the dark noise spectrum from the spectrum of each discharge condition.
7. The spectral bands and lines should thereafter be carefully analyzed for identification of the corresponding plasma species. For the identification, appropriate databases and references (e.g., [20–22]) can be employed. It should be ensured that the resolution of the spectrometer is taken into consideration for assigning the peaks to chemical species.
8. If the spectrometer is calibrated against a reference light source, use the multiplication factors for each wavelength to convert the data from arbitrary units to energy intensity. The spectra can thereafter be used for further processing and analysis, including line-ratio calculations or fitting for obtaining the gas temperatures.

4 Operator Safety

While cold plasma is a new technology to the food industry and there is limited information about its impact on the operator and lab personnel, the following are among the considerations for the operating cold plasma systems:

- Considering the concentration of ozone and reactive species, the system is preferred to be placed in a place with good ventilation such as cabinets equipped with laboratory hoods and carbon filters.
- The operator is expected to use an appropriate gas mask to be protected against ozone and reactive species.
- Due to the emission of UV radiation from some plasma sources, unprotected observation of the in-process sample for a long time

is not recommended. The operator may use a suitable safety goggle.

- As plasma systems usually involve high-voltage electricity, safety consideration regarding the use of electricity is needed.

5 Notes

1. If ethanol wipes are used, the traces of ethanol vapor can alter the plasma chemistry.
2. Remove the produce that shows deviation to achieve uniformity in size and desired developmental stage, and discard the produce that shows any surface damage such as bruising.
3. Sample should be dried in a biosafety cabinet or by the use of other appropriate methods.
4. This entire procedure should occur in the biosafety cabinet.
5. If the treatment is to be carried out using a modified gas atmosphere, the sterile packages have to be flushed with the gas mixture before sealing. For this, the gas must be passed through an appropriate filter (0.22 μm recommended).
6. There are variety of culture media, and the suitable one should be choose based on the aim of the test and the reference.
7. The time and the temperature can be varied based on the references.
8. For DBD systems operating at higher frequencies, appropriate kind of filtering of the signal using resistor-capacitor network will be required rather than just resistive division.
9. It is to be noted that as per theory of capacitors in series, the charge would remain the same in C_m and the DBD.

6 Conclusions

Cold plasma induced through gas ionization is being increasingly evaluated for practical applications involving decontamination of a variety of food products. A great deal of success has been observed in decontamination of fruits and vegetables using dielectric barrier discharges (DBDs). The experiments often involve the electrical characterization of the plasma using the Lissajous curve method which treats the DBD as a capacitor. The reactive species formed during the discharge are often detected using optical emission spectroscopy using fiber optic probes. The microbial inactivation efficacy of plasma is evaluated using various approaches involving either a surface treatment approach or whole product decontamination approach.

References

1. Misra NN, Martynenko A, Chemat F, Paniwnyk L, Barba FJ, Jambrak AR (2018) Thermodynamics, transport phenomena, and electrochemistry of external field-assisted non-thermal food technologies. Crit Rev Food Sci Nutr 58(11):1832–1863. https://doi.org/10.1080/10408398.2017.1287660
2. Mahnot NK, Siyu L-P, Wan Z, Keener KM, Misra NN (2020) In-package cold plasma decontamination of fresh-cut carrots: microbial and quality aspects. J Phys D Appl Phys 53(15):154002. https://doi.org/10.1088/1361-6463/ab6cd3
3. Misra NN, Yepez X, Xu L, Keener K (2019) In-package cold plasma technologies. J Food Eng 244:21–31. https://doi.org/10.1016/j.jfoodeng.2018.09.019
4. Moutiq R, Misra NN, Mendonca A, Keener K (2020) In-package decontamination of chicken breast using cold plasma technology: microbial, quality and storage studies. Meat Sci 159:107942. https://doi.org/10.1016/j.meatsci.2019.107942
5. Misra NN, Jo C (2017) Applications of cold plasma technology for microbiological safety in meat industry. Trends Food Sci Technol 64: 74–86. https://doi.org/10.1016/j.tifs.2017.04.005
6. Gavahian M, Chu YH, Jo C (2019) Prospective applications of cold plasma for processing poultry products: benefits, effects on quality attributes, and limitations. Compr Rev Food Sci Food Saf 18(4):1292–1309. https://doi.org/10.1111/1541-4337.12460
7. Wan Z, Misra NN, Li G, Keener KM (2021) High voltage atmospheric cold plasma treatment of Listeria innocua and Escherichia coli K-12 on Queso Fresco (fresh cheese). LWT Food Sci Technol 146:111406. https://doi.org/10.1016/j.lwt.2021.111406
8. Thomas-Popo E, Mendonça A, Misra NN, Little A, Wan Z, Moutiq R, Coleman S, Keener K (2019) Inactivation of Shiga-toxin-producing Escherichia coli, Salmonella enterica and natural microflora on tempered wheat grains by atmospheric cold plasma. Food Control 104:231–239. https://doi.org/10.1016/j.foodcont.2019.04.025
9. Misra NN, Kaur S, Tiwari BK, Kaur A, Singh N, Cullen PJ (2015) Atmospheric pressure cold plasma (ACP) treatment of wheat flour. Food Hydrocoll 44:115–121. https://doi.org/10.1016/j.foodhyd.2014.08.019
10. Mahnot NK, Mahanta CL, Farkas BE, Keener KM, Misra NN (2019) Atmospheric cold plasma inactivation of Escherichia coli and Listeria monocytogenes in tender coconut water: inoculation and accelerated shelf-life studies. Food Control 106:106678. https://doi.org/10.1016/j.foodcont.2019.06.004
11. Mahnot NK, Mahanta CL, Keener KM, Misra NN (2019) Strategy to achieve a 5-log Salmonella inactivation in tender coconut water using high voltage atmospheric cold plasma (HVACP). Food Chem 284:303–311. https://doi.org/10.1016/j.foodchem.2019.01.084
12. Misra NN, Keener KM, Bourke P, Cullen PJ (2015) Generation of in-package cold plasma and efficacy assessment using methylene blue. Plasma Chem Plasma Process 35(6): 1043–1056. https://doi.org/10.1007/s11090-015-9638-5
13. Gavahian M, Chu Y-H, Khaneghah AM, Barba FJ, Misra NN (2018) A critical analysis of the cold plasma induced lipid oxidation in foods. Trends Food Sci Technol 77:32–41. https://doi.org/10.1016/j.tifs.2018.04.009
14. Misra NN, Schlüter O, Cullen PJ (2016) Cold plasma in food and agriculture: fundamentals and applications. Academic Press, Elsevier, London
15. Bhullar MS, Monge-Brenes A, Perry B, Overdiep J, Nabwiire L, Shaw A (2021) Determining the potential food safety risks associated with dropped produce on floor surfaces in grocery stores. J Food Prot 84(2):315–320. https://doi.org/10.4315/JFP-20-136
16. Ziuzina D, Patil S, Cullen PJ, Keener KM, Bourke P (2014) Atmospheric cold plasma inactivation of Escherichia coli, Salmonella enterica serovar Typhimurium and Listeria monocytogenes inoculated on fresh produce. Food Microbiol 42:109–116. https://doi.org/10.1016/j.fm.2014.02.007
17. Charles MT, Arul J, Charlebois D, Yaganza E-S, Rolland D, Roussel D, Merisier MJ (2016) Postharvest UV-C treatment of tomato fruits: changes in simple sugars and organic acids contents during storage. LWT Food Sci Technol 65:557–564. https://doi.org/10.1016/j.lwt.2015.08.055
18. Jagadeesh SL, Charles MT, Gariepy Y, Goyette B, Raghavan GSV, Vigneault C (2009) Influence of postharvest UV-C hormesis on the bioactive components of tomato during post-treatment handling. Food Bioprocess Technol 4(8):1463–1472. https://doi.org/10.1007/s11947-009-0259-y

19. Moutiq R, Pankaj SK, Wan Z, Mendonca A, Keener K, Misra NN (2020) Atmospheric pressure cold plasma as a potential technology to degrade carbamate residues in water. Plasma Chem Plasma Process 40(5):1291–1309. https://doi.org/10.1007/s11090-020-10093-z
20. Gaydon A, Pearse R (1976) The identification of molecular spectra. Chapman and Hall, London
21. Herron JT, Green DS (2001) Chemical kinetics database and predictive schemes for nonthermal humid air plasma chemistry. Part II. Neutral species reactions. Plasma Chem Plasma Process 21(3):459–481
22. Laux C, Spence T, Kruger C, Zare R (2003) Optical diagnostics of atmospheric pressure air plasmas. Plasma Sources Sci Technol 12:125

Chapter 3

Plasma-Activated Water: Methods and Protocols in Food Processing Applications

Manreet Bhullar, Mohsen Gavahian, and N. N. Misra

Abstract

Plasma-activated water is finding increasing interest among food and agricultural scientists for advantaged applications. The chemistry of PAW involves formation of an array of reactive oxygen and nitrogen species, rendering their quantification difficult. However, several established chemical assays exist for quantification of the long-lived and major reactive species in PAW, such as the nitrate, nitrite, ammonium, and peroxide ions. For rapid analysis of many samples, absorption spectroscopy can also be employed, provided there are no other interfering substances. In addition, empirical physical chemistry approaches can also be employed for obtaining an overall picture of the reactiveness of the activated water using pH, ORP, and electrical conductivity. This chapter largely describes protocols for chemical characterization of PAW and touches upon microbial enumeration of foods treated using PAW.

Key words Cold plasma, PAW, Decontamination, ROS, RNS

1 Introduction: Cold Plasma Interactions with Water

Over the last two decades, there has been a significant surge in the research to apply gas plasma discharges for industrial applications, including electronics, polymers, medical, agriculture, and food processing. In recent times, the interest in plasma-liquid interactions has also increased significantly, as the liquid-chemistry changes significantly on exposure to plasma chemical species. Particularly, the interaction of ionized gases with water results in unique chemical properties, and the water is said to have become chemically active. The process results in plasma-activated water (PAW), which is the topic of the present chapter. PAW is being actively explored for disinfection and decontamination applications. PAW has also been reported to enhance the seed germination and seedling growth in many species of plants [1, 2]. These have direct consequences for the agriculture and malting industry [3–5]. The unique properties of PAW result from the reactive oxygen

Mohsen Gavahian (ed.), *Emerging Food Processing Technologies*,
Methods and Protocols in Food Science, https://doi.org/10.1007/978-1-0716-2136-3_3,

species (ROS) and the reactive nitrogen species (RNS) produced in water post-diffusion of the ionized gas. However, the production of these ROS and RNS invariably requires that the parent N_2 and O_2 molecules are present in the gas. The N_xO_y species formed in a N_2 and O_2 containing gas upon ionization results in the formation of HNO_x type of RNS on contact with water. These species form via the following schemes, R1–R4 [6, 7]:

$NO_2 + OH \rightarrow HNO_3$	(R1)
$NO_2 + HO_2 \leftrightarrow HO_2NO_2$	(R2)
$NO + O_2^- \rightarrow ONOO^-$	(R3)
$NO_2^- \rightarrow ONOO^- + H_2O$	(R4)

In addition, the ROS formed in the gas upon diffusion into water result in the formation of peroxides, hydroxyl radical, singlet oxygen, and dissolved ozone via the following reaction schemes, R5–R8:

$O_3 + H_2O \rightarrow H_2O_2 + O_2$	(R5)
$H_2O_2 \leftrightarrow HO_2^- + H^+$	(R6)
$O_2 + e^- \rightarrow 2O\bullet + e^-$	(R7)
$H_2O + e^- \rightarrow H\bullet + HO\bullet + e^-$	(R8)

A specific aim of the research in PAW is to tailor the underlying chemistry to suit the targeted application. Due to the complexity of the chemistry, involving gas phase ionization, diffusion, water-phase reaction, dissociation, and reaction with the substrate molecules, a variety of approaches are required for this purpose. This chapter focuses on the protocols of common assays and analyses for quantitative analysis of the reactive species in PAW. In addition, the chapter briefly elaborates the procedure for microbial inactivation assessment using PAW, being a topic of high interest to food scientists.

2 Materials

2.1 Chemicals and Equipment

Nitrite standard (0.1 M nitrite ion standard solution), Griess reagent, titanium oxysulfate sulfuric acid (high purity), analytical grade sodium nitrate, analytical grade sodium nitrite.

UV-Visible spectrophotometer, pH meter, ORP meter, cuvettes, nitrate-specific ion-selective electrode (ISE).

2.2 Water Samples

To induce reactive species in plasma from an external gas ionization source, it is recommended that deionized water be used, which typically has a resistivity of ~18.2 MΩ cm. Note that the resistivity (or conductivity as an alternate measure) of water is associated with the concentration of cations and anions. Further the entire stock of water intended for PAW generation should be stored under ambient conditions for sufficient time (say, 1 day) prior to exposure to electrical discharge. The storage would enable the concentration of dissolved gases in the water to equilibrate with the air, thereby ensuring that their concentrations are uniform for all experiments [8].

3 Methods

3.1 PAW Generation

Generally, two major methods of PAW production are available: (1) the interactions of liquid with the ionized gas and (2) direct generation of plasma in liquid. In this sense, the procedure provided by the plasma equipment supplier should be followed. For example, in terms of the first method, an appropriate volume of the liquid should be placed in the discharge area of plasma gas for an appropriate period of time [9].

The way that water contacts with electrical plasma generated via gas ionization affects the chemistry of PAW. The contact may be realized by means of a preexisting bubble, as a gas discharge above a liquid surface, or when the water is in the form of water droplets or sprays (*see* **Note 1**) [10]. Schematic representations of typical systems that are used for PAW generation can be found in Fig. 1.

3.2 Study of PAW Chemistry

The study of PAW chemistry involves three distinct aspects [1]:

1. Study of the active species formed in the gaseous discharge.
2. Investigation of the reaction of the gaseous species formed.
3. The transfer and fate of the gaseous reactive species into water.

Our focus in this chapter remains on the last aspect, wherein we describe the physicochemical approaches for such investigation. In order to obtain an overall picture of the extent of formation of ions, we typically employ measurement of the pH and electrical conductivity. For a more detailed analysis of the specific ions of interest, we seek individual assays for each ion, using either electrochemical measurements or spectrophotometric (or colorimetric) approaches. The protocols for these analyses follow hereunder.

3.2.1 Electrical Conductivity (EC)

The electrical conductivity (reported in μS/cm) of water is a measure to the capacity of water to conduct electrical current. This parameter is related to the concentration of salts dissolved in water. In case of PAW, the electrical conductivity provides an empirical estimate of the positively and negatively charged ions dissolved in water. The electrical conductivity of PAW can be determined by

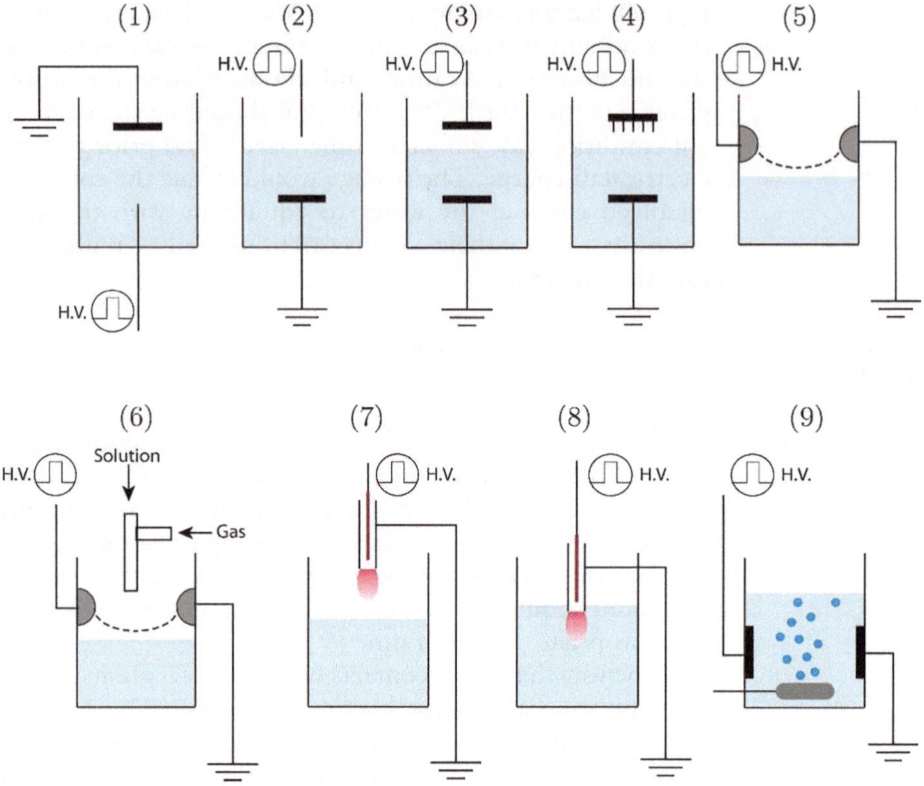

Fig. 1 Common systems for producing PAW. The type of discharge and the mode of introduction of reactive species into the liquid dictate the plasma chemistry. H.V. refers to high voltage. (Based on Zhou et al. [10], with modifications)

placing the probe of a portable EC meter and recording the reading (e.g., the PCSTestr 35 multiparameter meter from Eutech Instruments). The EC meter has two tiny electrodes separated by a known distance and dipped into the water sample. An electric potential is applied across these plates. The resulting current is measured by the meter and corresponds to the electrical conductivity.

3.2.2 pH and Oxidation-Reduction Potential (ORP)

ORP is an indicator of the overall level of reactive oxygen species. A higher measure of ORP implies a higher oxidation capacity of the PAW, which in turn is associated with microbial inactivation via quenching of the electrons from membranes of cell. Similarly, greater changes in the pH can be considered as higher level of water activation by plasma. pH and ORP of PAW can be measured by a pH-ORP meter [11]. In this regard, a volume of the sample suggested by the pH meter supplier should be placed in contact with the pH meter and ORP meter probe (*see* **Notes 2** and **3**).

3.2.3 Estimation of Nitrate (NO_3^-) Concentration

The concentration of nitrate in PAW can be estimated using a nitrate ion selective electrode or using chemical methods. In order to use the nitrate ion selective electrode, a calibration curve should be prepared to correlate the potential developed across the electrode (in mV) against the nitrate ion concentration.

One can prepare the calibration curve for nitrate analysis by performing serial dilution of a standard nitrate solution, say, with 100 ppm nitrate-N (i.e., $100 \times 4.43 = 443$ ppm of nitrate, OR $443/62.0049 = 7.145$ mM). Before each measurement, a nitrate-specific ionic strength adjusting buffer solution is always added at ~2% level to all samples and standards to create a uniform background ionic strength. This will enable to achieve reproducible measurements. Following precautions must be noted for the analysis:

1. Extreme care must be taken to avoid the sensing membrane from getting damaged or contaminated during the preparation of the electrode.
2. Care must be taken that samples and standard should be near the same temperature. A 1 °C difference in temperature for a 10^{-3} M nitrate solution could result in ~1.5% error.
3. The filling solution inside the electrode could crystallize if evaporates; thus, this should be avoided.

An alternative to ISE method for determination of nitrate concentration is via spectroscopic method (colorimetry). This method is described in *Standard Methods for the Examination of Water and Wastewater*, 22nd edition, SM 4500 NO^{3-} E (1997/2011) [12], and International Standard Organization ISO Standard No.7890-1 [13]. The 2,6-dimethylphenol (DMP) reagent used in this method is readily available from several suppliers (e.g., Nitrate Spectroquant® Analytical Test Kit from Merck) and has been employed for determination of nitrates in PAW [14].

3.2.4 Estimation of Nitrites (NO_2^-) Concentration

The nitrite (NO_2^-) ion concentration can be measured using a spectrophotometric method involving the Griess reagent. The Griess reagent contains naphthylethylenediamine dihydrochloride suspended in water and sulfanilamide in phosphoric acid. During reaction, sulfanilic acid is quantitatively converted to a diazonium salt by reaction with nitrite in acid solution. The diazonium salt is then coupled to *N*-(1-naphthyl)ethylenediamine, forming an azo dye that can be spectrophotometrically quantified based on its absorbance at 548 nm (Fig. 2):

For ion estimation, add 50 μL of Griess' reagent to 50 μL of the PAW and control on a microtiter plate. After 30 min of incubation, record the absorbance at ~548 nm using a 96 well microtiter plate reader. Alternatively, the volumes can be scaled appropriately for traditional 1 cm path length cuvettes are to be employed. However,

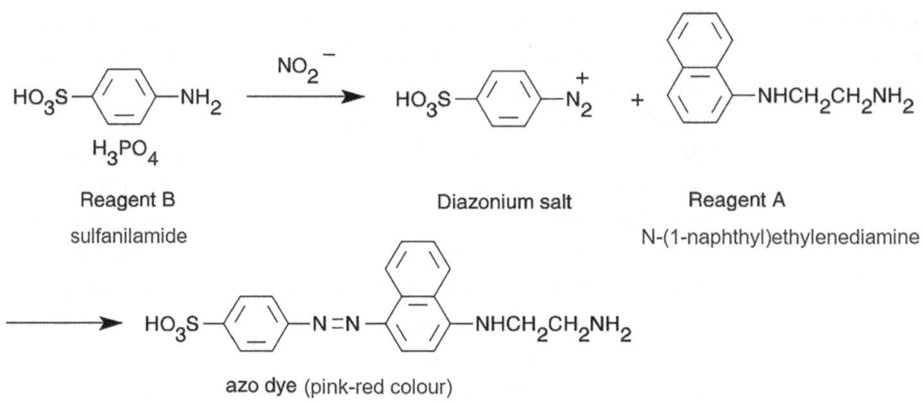

Fig. 2 Principle of nitrite quantitation using the Griess reaction

the microtiter plate method allows to quickly analyze a large number of samples. A range of sodium nitrite ($NaNO_2$) solutions of known concentrations (0, 40, 50, 60, 80, 100, 200, 300 μM) are to be prepared, and their post-incubation absorbances are to be recorded to build a nitrite calibration curve. This curve can subsequently be used for identification of the nitrite levels in the PAW.

3.2.5 Determination of Peroxide (O_2^{2-}) Content

The H_2O_2 concentrations in PAW can be determined via the titanium oxysulfate ($TiOSO_4$) colorimetric method. The peroxide in the water reacts with titanium sulfonate reagent to produce pertitanic acid, which is yellow in color with an absorption maximum near 405 nm. The reaction is effectively instantaneous, and the yellow-colored complexes are stable for ca. 6 h [15].

For quantification of peroxide in PAW, add a total of 10 μL acidic $TiOSO_4$ solution to 100 μL of PAW and control. After 10 min of incubation, record the absorbance on a spectrophotometer at 405 nm wavelength. For preparation of standard curve, record the absorbance of known H_2O_2 concentration solutions through identical process. As an example, to prepare H_2O_2 calibration curve, 30% hydrogen peroxide standard solution (e.g., Perhydrol® for analysis available under tradename EMSURE®) can be diluted into a concentration range of 0, 2×10^{-4}%, 3×10^{-4}%, 5×10^{-4}%, 1×10^{-3}%, 2×10^{-3}%, 3×10^{-3}%, and 5×10^{-3}% (1% = 0.3263 M). The unknown concentrations can then be determined from the calibration curve.

3.2.6 Determination of Ammonium Ions (NH_4^+)

The concentration of ammonium ions in PAW can be quantified using Nessler reagent. Nessler reagent reacts with ammonia in basic solution to form an orange-colored complex with an absorption maximum at 400 nm that can be recorded via a spectrophotometer [16]. The Nessler's reagent is readily available from major chemical suppliers. Alternatively, it can be prepared using anhydrous mercuric iodide and anhydrous potassium iodide or by dissolving potassium tetraiodomercurate in potassium hydroxide. For quantitative determination, add 2 mL of Nessler reagent to 50 mL of PAW, and record its absorbance after 10 min of incubation. To prevent any thermal decomposition side effects on quantification, the PAW samples are recommended to be stored at room temperature prior to mixing with Nessler reagent [17]. For conversion of absorption value to concentration, prepare a calibration curve using dilutions of ammonium chloride solution in water to obtain the linear relationship (e.g., a range between 0 and 277 μmol/L is reported in literature by Judee et al. [17]).

3.2.7 Reactive Species Estimation via Multiwavelength Transmission Spectroscopy

The concentration of multiple reactive species in plasma-activated water can simultaneously be estimated via light transmission spectroscopy and numerical deconvolution methods. This method relies on the UV-Vis light absorption patterns of reactive species formed in PAW. Following protocol is recommended for this realizing method:

1. Use a high-resolution spectrometer with a fixed spectral resolution (0.1–0.2 nm is recommended) and suitable scan speed. Use a 1 cm path length quartz cuvette for all light transmission measurements to ensure a high transmittance over the whole UV-Vis wavelength range. Turn on the spectrometer at least 20 min prior to measurements to avoid fluctuations in light intensity due to temperature variations during warmup.
2. Measured the transmittance of the control (T_0) and PAW (T) with empty quartz cuvettes as reference. The absorbance is then calculated as:

   $$A(\lambda) = -\log\left(T_0/T\right)$$

   It should be ensured that the time to measurement after plasma activation should be same for all samples and be as quick as possible. Reactive species such as OH• and peroxynitrite ($ONOO^-$) have a lifetime of less than 1 s, and these could be expected to have extinguished by the time of measurement [8]. Thus, nitrate, nitrite, and peroxides can be expected to be present in PAW and quantified by this method.
3. The concentration of reactive species in PAW can be determined from the spectra using the Beer-Lambert law:

   $$A(\lambda) = \varepsilon(\lambda) l c_x$$

   where ε is the molar absorptivity of the chemical species at a certain wavelength λ, l is the optical path length, and c_x is the concentration of reactive species.

   When several chemical species, c_1, c_2, c_3, …, c_n contribute to the absorption spectrum, the absorbance per unit path length is given by:

$$\frac{A(\lambda)}{l} = \sum_{i=1}^{n} \varepsilon_i(\lambda) \cdot c_i$$

The concentration of each species can then be determined mathematically if the ratios between absorptivity are considerably different among the reactive species.

4. The concentration of four model species, H_2O_2, NO_2^-, NO_3^-, and O_2, can be determined by solving the following matrix equations:

$$\frac{1}{l} \cdot \begin{bmatrix} A(\lambda_1) \\ A(\lambda_2) \\ A(\lambda_3) \\ A(\lambda_4) \end{bmatrix} = \begin{bmatrix} \varepsilon_1(\lambda_1) & \varepsilon_2(\lambda_1) & \varepsilon_3(\lambda_1) & \varepsilon_4(\lambda_1) \\ \varepsilon_1(\lambda_2) & \varepsilon_2(\lambda_2) & \varepsilon_3(\lambda_2) & \varepsilon_4(\lambda_2) \\ \varepsilon_1(\lambda_3) & \varepsilon_2(\lambda_3) & \varepsilon_3(\lambda_3) & \varepsilon_4(\lambda_3) \\ \varepsilon_1(\lambda_4) & \varepsilon_2(\lambda_4) & \varepsilon_3(\lambda_4) & \varepsilon_4(\lambda_4) \end{bmatrix} \begin{bmatrix} c_1 \\ c_2 \\ c_3 \\ c_4 \end{bmatrix}$$

The left-hand side of the above equation is known from experimental data. Thus, one could solve for the concentrations using matrix inversion:

$$\begin{bmatrix} c_1 \\ c_2 \\ c_3 \\ c_4 \end{bmatrix} = \frac{1}{l} \begin{bmatrix} \varepsilon_1(\lambda_1) & \varepsilon_2(\lambda_1) & \varepsilon_3(\lambda_1) & \varepsilon_4(\lambda_1) \\ \varepsilon_1(\lambda_2) & \varepsilon_2(\lambda_2) & \varepsilon_3(\lambda_2) & \varepsilon_4(\lambda_2) \\ \varepsilon_1(\lambda_3) & \varepsilon_2(\lambda_3) & \varepsilon_3(\lambda_3) & \varepsilon_4(\lambda_3) \\ \varepsilon_1(\lambda_4) & \varepsilon_2(\lambda_4) & \varepsilon_3(\lambda_4) & \varepsilon_4(\lambda_4) \end{bmatrix}^{-1} \begin{bmatrix} A(\lambda_1) \\ A(\lambda_2) \\ A(\lambda_3) \\ A(\lambda_4) \end{bmatrix}$$

Note that only four distinct wavelengths have been chosen for the calculation to satisfy the criteria of a unique solution and for simplicity. Reactive species concentration in PAW using this method was successfully employed by Oh et al. [8], wherein they have also experimentally retrieved the absorption coefficient data. One could also select multiple wavelengths as long as the absorptivity data is available over that wavelength window. For such a method, iterative optimization routines can be employed.

3.2.8 Overview of the Equipment That Can Be Used for RONS-Related Analysis in PAW

The interactions of plasma and liquids, such as water, involve many direct reactions at the interface of gas-liquid and indirect cascade items in the liquid, producing a liquid that has a mixture of plasma-generated chemicals such as reactive species. These compounds can be classified into short- and long-lived groups. Short-lived reactive species in PAW include singlet oxygen, $^1O_2$, hydroxyl radicals, •OH, atomic oxygen, O, and the superoxide radical, $O_2•^-$, while long-lived include H_2O_2, O_3, and secondary species such as NO_2^- and NO_3^-. Qualitative and quantitative analyses of these species need to be performed with several considerations (*see* **Notes 4** and **5**). Optical spectrophotometry, fluoro-luminescence spectroscopy, electron spin resonance (ESR) spectroscopy, chromatography, and mass spectrometry are among the common techniques that can be used for such analyses. The details for operating such equipment are specific to the equipment and provided by the supplier [10].

3.3 Decontamination of Food with PAW

There are several protocols to follow for microbiological assays, depending on the type of microbiological tests (e.g., measuring total bacterial count, yeast, mold, or a specific type of bacterium, such as *Escherichia coli*). Among these methods, the 3 M Petrifilm method provides a quick and simple approach. Hence, a protocol based on such an approach to measure the aerobic bacteria before and after plasma treatment is provided.

3.3.1 Preparation of Sample

1. Sterile diluents: Different types of aqueous chemical solutions can be used for dilution purposes, including Butterfield's phosphate buffer, 0.1% peptone water, peptone salt diluent, buffered peptone water, saline solution (0.85–0.90%), or distilled water (*see* **Note 6**).
2. Per standard, the sample should be homogenized (*see* **Notes 7** and **8**).

3.3.2 Plating

1. Petrifilm plate should be placed on a flat, level surface.
2. The edge of the top film should be lifted for dispensing the sample suspension (1 mL) perpendicularly on the center of the bottom film using a precise pipette.
3. The top film should be rolled down to avoid trapping of any bubbles.
4. A flat spreader should be used to provide a consistent distribution of samples over the plate (*see* **Note 9**).
5. The plate should be kept in a place without moving for appropriate gel formation while using the spreader for 1 min.

3.3.3 Incubation

Plates should be incubated horizontally while keeping the clear side up at a temperature below 40 °C.

3.3.4 Interpretation

An illuminated magnifier or colony counter can be used to determine the microbial count. All colonies should be counted regardless of size and color. Then, the total aerobic bacteria count can be calculated based on the dilution performed during this test.

4 Notes

1. The type of plasma source, treatment time, carrier gas, and the gap between plasma discharge area and the water surface are among crucial parameters that can affect the chemistry of PAW.
2. The temperature of the water during measurement should be accounted for in the calculations, where applicable.

3. It is recommended that this analysis be performed immediately after PAW production as this value is subject to changes during storage period.
4. Plasma and liquid interactions are dynamic which can make the diagnostics challenging, specifically, when it comes to short-lived reactive species of PAW. Therefore, advanced analytical approaches must be leveraged, and the progress in the analytical methods should be kept in mind.
5. Plasma chemistry occur on sub-second scales (ns, μs, and ms). Therefore, such analysis should be performed online or as soon as possible.
6. Diluents containing bisulfite, citrate, or thiosulfate are not recommended in this method as these chemicals as the growth of microorganism might be limited in presence of such materials.
7. 1N NaOH can be used to adjust the sample pH to above 5 to obtain the optimum growth and recovery of microorganisms in acidic samples.
8. These experiments should be performed in sterilized conditions such as using a microbiological laminar flow cabinet.
9. The plate should be kept in a place without moving for appropriate gel forming while using the spreader for 1 min.
10. Incubation temperature and time varies among references.

5 Conclusions

Plasma-activated water is finding increasing interest among food and agricultural scientists for advantaged applications. The chemistry of PAW involves a variety of reactive oxygen and nitrogen species that form and exist over a variety of timescales (ns, μs, and ms) rendering their quantification quite difficult. However, several established chemical assays exist for quantification of the long-lived and major reactive species in PAW, such as the nitrate, nitrite, ammonium, and peroxide ions. For rapid analysis of many samples, absorption spectroscopy can also be employed, provided there are no other interfering substances. In addition, empirical physical chemistry approaches can also be employed for obtaining an overall picture of the reactiveness of the activated water using pH, ORP, and electrical conductivity. Depending on the application in hand, it may be necessary to employ advanced instrumental analysis methods, such as mass spectrometric analysis or electron spin resonance (ESR).

References

1. Bradu C, Kutasi K, Magureanu M, Puač N, Živković S (2020) Reactive nitrogen species in plasma-activated water: generation, chemistry and application in agriculture. J Phys D Appl Phys 53(22):223001. https://doi.org/10.1088/1361-6463/ab795a
2. Misra NN, Schlüter O, Cullen PJ (2016) Cold plasma in food and agriculture: fundamentals and applications. Academic Press, Elsevier, London
3. Thirumdas R, Kothakota A, Annapure U, Siliveru K, Blundell R, Gatt R, Valdramidis VP (2018) Plasma activated water (PAW): chemistry, physico-chemical properties, applications in food and agriculture. Trends Food Sci Technol 77:21–31. https://doi.org/10.1016/j.tifs.2018.05.007
4. Puač N, Gherardi M, Shiratani M (2018) Plasma agriculture: a rapidly emerging field. Plasma Process Polym 15(2):1700174. https://doi.org/10.1002/ppap.201700174
5. Ohta T (2016) Plasma in agriculture. In: Misra NN, Schlueter O, Cullen PJ (eds) Cold plasma in food and agriculture, vol 8. Elsevier, London, pp 205–221. https://doi.org/10.1016/b978-0-12-801365-6.00008-1
6. Arjunan KP, Obrusník A, Jones BT, Zajíčková L, Ptasinska S (2016) Effect of additive oxygen on the reactive species profile and microbicidal property of a helium atmospheric pressure plasma jet. Plasma Process Polym 13:1089. https://doi.org/10.1002/ppap.201600058
7. Wan Z, Misra NN, Li G, Keener KM (2021) High voltage atmospheric cold plasma treatment of Listeria innocua and Escherichia coli K-12 on Queso Fresco (fresh cheese). LWT Food Sci Technol 146:111406. https://doi.org/10.1016/j.lwt.2021.111406
8. Oh J-S, Szili EJ, Ogawa K, Short RD, Ito M, Furuta H, Hatta A (2018) UV–vis spectroscopy study of plasma-activated water: dependence of the chemical composition on plasma exposure time and treatment distance. Jpn J Appl Phys 57(1):0102B0109. https://doi.org/10.7567/jjap.57.0102b9
9. Gavahian M, Peng HJ, Chu YH (2019) Efficacy of cold plasma in producing Salmonella-free duck eggs: effects on physical characteristics, lipid oxidation, and fatty acid profile. J Food Sci Technol 56(12):5271–5281. https://doi.org/10.1007/s13197-019-03996-z
10. Zhou R, Zhou R, Wang P, Xian Y, Mai-Prochnow A, Lu X, Cullen PJ, Ostrikov K, Bazaka K (2020) Plasma-activated water: generation, origin of reactive species and biological applications. J Phys D Appl Phys 53(30):303001. https://doi.org/10.1088/1361-6463/ab81cf
11. Gavahian M, Sheu FH, Tsai MJ, Chu YH (2019) The effects of dielectric barrier discharge plasma gas and plasma-activated water on texture, color, and bacterial characteristics of shiitake mushroom. J Food Process Preserv 44(1):e14316. https://doi.org/10.1111/jfpp.14316
12. Eaton E, Baird R, Rice E (2012) Standard methods for the examination of water and wastewater, 22nd edn. APHA, AWWA, WEF, Washington, DC
13. ISO (1986) Water quality - determination of nitrate -Part 1: 2,6-Dimethylphenol spectrometric method, vol 7890-1. International Standards Organization, Geneva
14. Lu P, Boehm D, Bourke P, Cullen PJ (2017) Achieving reactive species specificity within plasma-activated water through selective generation using air spark and glow discharges. Plasma Process Polym 14(8):1600207. https://doi.org/10.1002/ppap.201600207
15. Eisenberg G (2002) Colorimetric determination of hydrogen peroxide. Ind Eng Chem Anal Ed 15(5):327–328. https://doi.org/10.1021/i560117a011
16. Crosby NT (1968) Determination of ammonia by the Nessler method in waters containing hydrazine. Analyst 93(1107):406–408
17. Judee F, Simon S, Bailly C, Dufour T (2018) Plasma-activation of tap water using DBD for agronomy applications: identification and quantification of long lifetime chemical species and production/consumption mechanisms. Water Res 133:47–59. https://doi.org/10.1016/j.watres.2017.12.035

Chapter 4

Ozone Processing of Foods: Methods and Procedures Related to Process Parameters

R. Pandiselvam, V. Prithviraj, Anjineyulu Kothakota, and Krishna Prabha

Abstract

Ozone technology in food processing technique had expanded largely over the years. The environment-friendly nature of the technology makes it a compelling option among consumers. The multiprocessing aspects of ozone technology make it effective against various microorganisms in fruits, meat, and vegetables and simultaneously performing water treatment. Ozone generation is carried out using pure oxygen, and it immediately decomposes to oxygen after treatment thereby leaving no toxic effects. A proper ozone treatment retains nutritional, sensory, and physiochemical properties of the products. This chapter intends to cover factors affecting ozone technology along with basic design of an ozone treatment system.

Key words Ozone technology, Equipment design, Emerging food processing technology, Food processing, Fruit and vegetable disinfection

1 Introduction

Ozone technology is one of the sustainable food processing methods which are explored in various food processing aspects. The technology can be seen in fruit and vegetable processing, grain processing, hydrocolloids, meat processing, sea food processing, and sanitation process. However, the high oxidizing potential of ozone is mostly used for cleansing and disinfection activities. Ozone molecule consists of three oxygen atoms in an isosceles triangle manner with bond angle of 116.8°. The name is derived from the word "ozein" meaning "to smell" since it also has a pungent odor [1].

The production of O_3 is carried out using high electrical discharge, electrochemical methods, or ultraviolet radiations that produce high energy capable of converting oxygen molecules into ozone. The high energy provided by the these methods split the diatomic oxygen into oxygen radicles which react with other diatomic oxygen to form ozone [2]. Commercially, electrical

Mohsen Gavahian (ed.), *Emerging Food Processing Technologies*,
Methods and Protocols in Food Science, https://doi.org/10.1007/978-1-0716-2136-3_4,

discharge method is used for the production; however, it was found to have low efficiency of 2–10% [1]. In addition, ozone gas has a half-life of 28 min in distilled water [3] which raise challenge in storing ozone without being converted to oxygen. Investigation conducted by Hirahara et al. [3] added citric acid (1 mmol/L) with distilled water thereby doubling the half-life of ozone molecules and thereby increasing the stability of the same.

The principle behind the action of ozone gas is its highly reactive nature that is employed for processing both direct and indirect methods. The direct mode employs ozone as nucleophilic or electrophilic agent where the former occurs due to shortage of reaction and later due to high electron density. However, the indirect mode of ozone reaction forms more complex species as intermediates those are highly reactive than ozone such as OH radicles. The reaction follows three major steps, viz., initiation, propagation, and termination starting from ozone and finally resulting in oxygen and water [4]. Similarly, these reactive species generated are responsible for causing various damages to microorganisms causing undesirable effects.

Furthermore, in application aspects, ozone is found to have a multipurpose role in processing. Washing the produces with ozonated water not only cleans the produce but also cuts down the wastewater processing cost due to its sanitization effects [5]. The microbial inactivation of ozone is found to be effective against *E. coli*, *Saccharomyces cerevisiae*, and *Salmonella* species during treatment. Also, gram-positive bacteria are more resistant to ozone due to the presence of high peptidoglycan in the cell wall, whereas gram-negative bacteria are easily destroyed by attaching at lipoprotein and lipopolysaccharide layers [5]. Ozone treatment is suitable for most of the fresh products, viz., root vegetables, apple, pear, kiwi, tomato, carrot extending to dried fig, raisins, and fruit juices. The application of ozone in gas phase was used for fungi inactivation and mycotoxin degradation especially in dried produces such as fig. Although ozone is also capable of providing pesticide degradation, there are chances of creating toxic by-products while doing so due to highly reactive ozone species.

As far as regulatory standards go, USFDA had given approval for using ozone as antimicrobial agent, and it is categorized under GRASS during the period of 1995–1997 (Generally Recognized as Safe) [6]. However, since the food processing industries all over the world are exploring various ozone-aided processes, more clear guidelines on regulatory standards would facilitate applied research in the field. In addition, consumers are more concerned about the various processing stages after harvest and demand minimally processed foods that are "farm fresh." Therefore, ozone technology and its advantages need to be communicated with the public for better acceptance of ozone-treated products.

Although ozone is widely used for food processing, significant exposure is toxic, primarily affecting respiratory organs even leading to fatality. Concentrations of 0.1 ppm are able to cause sharp irritation in the nose and throat which even leads to loss of vision at increased concentration about 0.5 ppm for 3–6 h exposure. The throat irritation gets severe at concentrations of 1–2 ppm causing cough, chest pain, and headache. Even higher level of exposure at 5–10 ppm drastically affects the function of the lungs which leads to fatality at 50 ppm concentration [1, 2].

1.1 Advantages of Ozone

Ozone technology has several inherent advantages which render it as a promising processing method. However, there are some limitations out of which most of the technical ones can be improved in future research. The various advantages and limitations according to various researchers [5, 7–12] are summarized below.

1. The commercial availability of ozone along with its high oxidization potential can be used for water treatment and sanitization applications.
2. Cleaning of fresh commodities in food industries is done using water; hence, due to the stable and soluble nature of ozone, it can be incorporated with water for acting against microorganisms and pesticide residues along with cleansing action of water.
3. Ozone can be made from oxygen and can be decomposed to same which is an environment-friendly gas.
4. Ozone is instantly made and decomposes very fast; thus, any malfunction of the system can be arrested by stopping the ozone generator.
5. Due to the highly reactive species, the oxidization and disinfection reactions are fast compared to other traditional methods.
6. Running cost of the system is for oxygen cylinders that supply pure O_2 and electrical power.
7. Heat is not produced during ozone treatment; therefore, thermal damage of sensitive produces is prevented.

1.2 Limitations of Ozone

1. Exposure to high concentration of ozone for long time can cause fatality.
2. Sanitization effect of ozone only lasts for short duration; therefore, chance of recontamination exists.
3. Leakage detection devices are needed at indoor usage.
4. On-site development is needed due to highly unstable nature.
5. Initial investment for the setup is high unlike chemical disinfectants that are bought directly.
6. Storage and transport are challenges. Therefore, generation equipment is always needed for ozone processing.

7. Although it is cost effective than other existing technologies, the efficiency of corona discharge method of commercial ozone production is highly inefficient.

This chapter discusses about the food processing-related aspects of ozone technology. Also, various extrinsic and intrinsic factors that are related to ozone processing are discussed in detail. These parameters and their effects need to be considered during ozone processing, and the same needs to be optimized for best performance of the system. Furthermore, equipment setup, methods, and precautions along with its advantages and limitations are also given for better understanding of the technology.

2 Ozone Processing Parameters

The generated ozone cannot be stored for a longer period at room temperature as it degrades spontaneously due to high oxidation potential (+2.07 V) of the ozone. It was found that numerous factors such as presence of other gaseous substances including nitrous oxide, hydrogen sulfide, sulfur dioxide or trioxide, oxide anions, or atomic oxygen resulting from electric discharges [13] affect the stability of the ozone. Depending on the concentration, the ozone in ozone-oxygen mixture has a stability less than 100–200 h after which it starts to decompose at room temperature [14].

The ozone treatment has a broad application in food industry, and the treatment conditions vary depending on the food products chosen. The ozone treatment has been successfully used in food industries for wastewater treatment, for drinking water disinfection, for extending shelf life of the processed food products without affecting the quality, and for improving the functional properties of the food products. In perishable commodities such as fruits and vegetables, the ozone is efficient in reducing mold and other microbial growth, surface decontamination, preventing germination, pesticide removal, mycotoxin removal, and ethylene removal without significantly affecting the quality attributes of the product. However, the efficacy of ozone treatment on the commodity is significantly influenced by certain extrinsic and intrinsic parameters and will be discussed in the following sections below.

2.1 Extrinsic Parameters

2.1.1 Concentration of Ozone

The concentration of ozone gas is expressed as g/m^3 or percent by weight and could be quantified using different methods. For aqueous form of ozone, both spectrophotometric and volumetric methods could be used of which the later one is labour intensive (e.g. Iodometric titration) are used. In the physicochemical method or spectrophotometric method of measurement, the ozone is allowed to stoichiometrically react with a particular compound such as in potassium iodide (in DDPD procedure) or indigo trisulfonate (in indigo procedure), and the transformation is evaluated as spectroscopic absorption under ultraviolet or visible light [15–17]. For quantifying gaseous ozone, the physical and photometric method (ozone analyzer) could be used which involves measuring the UV light absorption by ozone molecules in the presence of spectroscopic source (e.g., mercury).

While generating ozone, the ozone concentration should not exceed a limit that results in self-destruction. A high concentration of ozone increases the rate of ozone degradation. For every food commodity, there is a threshold ozone concentration and exposure time above which the product damage might arise [18]. The optimum concentration of ozone required is highly dependent on medium in which it is applied. The ozone is highly efficient at low concentration with less exposure time to eliminate microbial contamination.

The ozone treatment is efficient for a broad spectrum of microorganisms [19]. Although gram-positive bacteria such as *Enterococcus faecalis*, *Staphylococcus aureus*, *Listeria monocytogenes*, and *Bacillus cereus* were found to be sensitive to ozone treatment, they are highly efficient in inactivating gram-negative coliforms. However, some other studies state that the gram-positive bacteria such as *Listeria monocytogenes* were more sensitive to aqueous ozone treatment than the gram-negative ones [13]. It is recognized that the higher content of peptidoglycan in their cell wall might be a reason for the resistance.

The food storage room requires large volume of lower concentration ozone containing air as high ozone concentration might result in surface discoloration in fruit and vegetables. However, a high concentration of ozone is required for treating wastewater or removing pesticides from the food commodity. An optimum amount of ozone concentration should be selected to achieve maximum inactivation level considering the fast dissipation of ozone due to factors that affect its efficacy. The concentration ozone should be regularly monitored in food industries to avoid health issues related to ozone exposure. According to occupational health and safety administration, the exposure to gaseous ozone is limited to a concentration of 0.1 part of ozone per million parts of air (ppm) over an 8 h shift [20].

2.1.2 Flow Rate

In aqueous treatment medium, a high ozone gas flow rate may result in development of large-sized bubbles and thereby less dissolution of ozone in water. However, lower flow rate results in small-sized bubbles leading to less inactivation rate due lower ozone concentration [4]. Keeping all other factors constant, the treatment efficacy was found to be increased with a smaller bubble size due to better mass transfer [7]. The size of bubbles in the ozonized water is crucial in studying the disinfection rate. The microbial inactivation rate and residual ozone in water were increased when the bubble size was reduced from 2.38 to 1.72 mm under constant ozone flow rate and given concentration [21]. For the inactivation of gram-negative microorganism such as *E. coli ATCC 25922*, an optimum flow rate of 0.12 L/min was chosen with an apparent *D* value of 2.69 min and a lag time of 5.47 min keeping other extrinsic parameters (temperature, exposure time, and concentration) constant [22].

It is found that flow rate of gaseous or aqueous ozone plays a significant role in determining the quality of treated products. In meat processing, a gas flow rate of >2 μg/L was effective against *Pseudomonas* spp. on reducing the count on contaminated beef [23]. However, the gas flow rate is highly depended on the concentration and state of ozone under study. A reduction in pathogen count was observed in contaminated beef carcasses when treated with aqueous ozone at a flow rate of 95 mg/L ozone concentration at 80 $lb/in.^2$ compared to plain water washing [24]. In fruit beverages such as orange juices, an increase in flow rate significantly affected the color stability. There was a significant increase in L^* values with a flow rate in the range of 0–0.25 L/min, whereas in values a^* and b^*, it was found to be reduced [25]. The ozone could be successfully applied on reducing infestation by insects such as *T. castaneum*, *O. surinamensis*, and *R. dominica*, with a concentration of 150 ppm ozone at a flow rate of 2 L/min. Finally, one has to ensure that the gas flow rate should be optimized by maximizing the solubility and minimizing the gaseous escape from the medium.

2.1.3 Physical State of Ozone

The ozone in gaseous state is found to be more stable than in aqueous medium. However, the ozone gas cannot be stored for a longer time as it decomposes back into oxygen atoms on storage [2]. The exposure time was found to be longer when the harvested product is treated immediately with gaseous ozone than ozone dissolved in aqueous medium. Thus ozone gas can be used at lower concentration (0.1–10 μL/L) on freshly harvested fruits and vegetables by adding to the storage atmospheres for extending shelf life without compromising the quality [16]. It was found that a gaseous ozone concentration as low as 0.2 ppm along with refrigerated condition (4 °C) was sufficient for extending shelf life of fresh-cut lettuce for 7 days. There was significantly lower microbial load (*E. coli*, *Pseudomonas*, yeast, and mold counts) and better color during the storage period [26]. However, many literatures state that gaseous ozone is a less effective antimicrobial agent than aqueous ozone and a higher relative humidity is required for ozone gas for increased antimicrobial properties [16, 27]. A higher relative humidity (90–95%) is required for ozone gas for the inactivation of microorganisms. If the humidity of gas reaches below 50%, least bactericidal properties were reported [28].

2.1.4 Temperature

The ozone is least stable with an increase in the temperature. The half-life of the ozone is also affected by the temperature as the decomposition of ozone to oxygen is faster when the temperature of medium is increased [5]. In aqueous medium, the stability ratio of the ozone increased with a decrease in temperature. A lower temperature resulted in low solubility of ozone in water and faster decomposition rate [29].

The temperature plays a significant role in enhancing the antimicrobial properties of ozone. Steenstrup and Floros [30] studied the effect of gaseous ozone at varying temperature (0–20 °C) concentration above 1000 ppm on *E. coli 0157:H7* in apple cider. With increase in temperature, the *D* value was found to be reduced with shorter lag time indicating that higher temperature significantly increased the rate of inactivation of the microorganism. Similarly, Wu et al. [31] also observed an increase in efficacy of ozone against infestation with increase in temperature from 10 to 40 °C. In contrast to these results, the effect of gaseous ozone on aerobic mesophilic and heterotrophic microorganisms and inoculated *Escherichia coli* in culture media and beef samples was demonstrated by [32]. There was a complete inactivation of *Escherichia coli* in culture media when treated with gaseous ozone (154×10^6 kg m^3) at refrigerated temperatures (0–4 °C). A similar study was conducted by [33] to elucidate the effect of gaseous ozone (10×10^{-6} kg O_3/h) exposure on *Escherichia coli* O157:H7-contaminated ground Hanwoo beef at refrigerated condition (4 °C) for 3 days. They successfully demonstrated the antimicrobial activity of ozone under refrigerated condition of meat products and provided a foundation for integrating ozone generator inside a refrigerator as part of extending the shelf life of perishable commodities. Thus, we could conclude that the ozone was equally found to be efficient at low and high temperatures. However, one has to keep in mind that the temperature has a significant role on stability and solubility of ozone.

2.1.5 Exposure Time

Ozone is a powerful oxidative biocide and requires less exposure time and concentration for microbial reduction as compared to antimicrobial such as chlorine and peracetic acid. In a study conducted by [18], ozone concentration of 1.4 mg/L, for 1, 3, and 5 min, was applied on fresh-cut onions to eliminate native microflora and quality retention during storage at 4 °C for 14 days. An increase in exposure time resulted in high log reduction of microbial load and residual levels of pesticides. In a similar study conducted by [34] on fresh-cut cabbage with the same flow rate and exposure times, an aqueous ozone treatment of 5 min was found to be an economical and viable method for eliminating aerobic bacteria, coliforms, and yeasts and removing pesticides, thereby maintaining the storage quality of fresh-cut cabbages. In meat processing, the combined use of ozone (0.6 ppm for 10 min) and freeze drying enhanced the shelf life up to 8 months without compromising sensory quality and microbial safety of the broiler chicken meat [35].

According to [36], an exposure time of 1 h at 5 ppm is required for reducing both coliform and *Staphylococcus aureus* populations on date fruits, as long exposure time is required for eliminating mesophilic microbes, yeasts, and molds. Among other microorganisms, yeast and molds are highly tolerant to ozone treatment and require prolonged treatment compared to other microflora. Even though a prolonged exposure of ozone towards microbes successfully eliminated them, the exposure time significantly affected the quality of the produce. The research conducted by [37] confirmed that the polyphenols are significantly affected with an increase in contact time. It was also found that there was a symptom of toxicity of certain cultivars of grapes upon prolonged exposure to ozone treatment above 20 min [38]. To conclude, the exposure time and concentration should be kept as low as possible enough to inactivate the microorganisms but still to ensure better quality of food commodities.

2.2 Intrinsic Parameters

2.2.1 pH

A change in pH of the medium will substantially contribute to the efficacy of ozone as the decomposition reaction of ozone varies at different pH. It was found that under acidic conditions (pH around 3.00–4.00), the decomposition rate of ozone is comparatively slow and thus more stable [16]. The ozone decomposition is a result of hydroxyl radical formation with increase in pH (up to pH value of 7). Above pH 9, initiation of peroxy radicals and hydroxide ion accelerates the decomposition. Consequently, studies proved that the antimicrobial activity of ozone is well exhibited when the food or aqueous medium has a lower pH [16, 39, 40].

The relationship between pH levels and effectiveness of gaseous ozone on the inactivation of *Escherichia coli* ATCC 25922 and NCTC 12900 strains in apple juice under constant flow rate (0.12 L/min), concentration (0.048 mg/min/mL), and exposure time (18 min) was studied by [41]. They found that the duration required for 5-log reduction was faster (4 min) at the lowest pH than at the highest pH (18 min) and described mathematically the relationship between time required to achieve 5-log reduction (t_{5d}) and pH for both strains. A study was conducted by [42] on Brazil nuts to evaluate the decomposition kinetics of aqueous media at different pHs. The study also investigated the capability of ozone in inactivating *Aspergillus flavus* and in determining changes in nut color and crude oil lipid profile. Though there was an initial increase in concentration of ozone dissolved in aqueous media with brazil nuts, the half life of ozone was found to be reduced with reduction in pH. However, the ozonation in acidic medium was proved to be the most effective method in controlling fungal species (*Aspergillus flavus*) without compromising the quality.

Aqueous ozone (1.5 ppm) in combination with propionate, acetate, or butyrate short-chain fatty acids (SCFAs) as well as citrate or oxalate buffer formulations that maintain an acidic pH against *S. aureus*, *S. enterica*, and *K. pneumonia* was demonstrated by [43]. There was a 4-log or greater reduction in CFUs posttreatment for all three species versus treatment with water alone hence confirming the fact that ozone is highly efficient in acidic medium.

2.2.2 Organic Matter

Organic matter and their presence are an important aspect which affects the efficacy of ozonation. Nonetheless, it is noteworthy that the type of organic matter that exists during ozone treatment is more significant than the amount present. Accordingly, Khadre and Yousef [44] suggested the undesirability of organic matter in water which is employed for food processing. Further, the authors stated that the solid matter on their reaction with ozone may produce several disagreeable by-products which will consequently degrade the shelf stability, sensorial properties, as well as safety of the final product [44]. Moreover, Güzel-Seydim et al. [45] investigated the efficiency of ozone in reducing microbial populations in the presence of food components. Sterile class C buffer, locust bean gum, soluble starch, sodium caseinate, and whipping cream were employed as biomaterials in the experiment. The results revealed that caseinate and whipping cream had most pronounced effect in protecting the bacterial spores of *B. stearothermophilus* and the vegetative cells of *E. coli* and *S. aureus* [45]. However, starch failed in providing protection to the bacterial population, while locust bean gum had an intermediate protection level. Therefore, the research indicated the bactericidal effect of food components which was in accordance with the statement by Restaino et al. [46]. Likewise, it has been observed that the antimicrobial efficiency of ozone was reduced due to the presence of organic matter and ascorbic acid in orange juice [47]. The reduced efficiency of ozone may be attributed to the high ozone demand of the solid material present in the food product. Similar results were obtained by Patil et al. [22] in which the organic matter significantly reduced the bactericidal activity of gaseous ozone. Furthermore, it is observed that the unfiltered orange juice took 15–18 min for microbial inactivation when compared with shorter inactivation times in juice with low pulp [22]. Further, the presence of several compounds like lipids, sugars, reducing agents (like ascorbic acid), fibers, carbohydrates, etc. will act as barriers against ozone and will decrease its oxidizing strength. Consequently, the by-products formed will have little or no antibacterial activity [7, 48]. Hence, while designing ozonation systems, it's important to consider solid matter content of the product to be treated as the organic load in it will decrease the antimicrobial effectiveness of ozone and also may deteriorate the organoleptic characteristics of the treated product.

3 Equipment in Ozonation System

The ozonation system for treating numerous food products involves the use of certain equipment, viz., oxygen concentrator, ozone generator, collector, ozone analyzer, and ozone destructor. Thus, in the process of ozone production, oxygen or atmospheric air enriched with oxygen will be fed to the generator by the concentrator. Furthermore, the generated ozone is transferred to the collector for treating the food produce, and thereafter the surplus gas will be broken down by the destructor. The following Fig. 1 is a diagrammatic representation of the ozonation system.

The subsequent section deals with the details of the equipment in ozonation system and their working.

3.1 Oxygen Concentrator

The instruments which are intended to provide oxygen enriched (more than 90%) feed gas to the ozone generator is termed as oxygen concentrators. Moreover, these equipment filter undesirable dust particles from air, dry it to a temperature of 4 °C which is below maximum desired dew point, and then supply feed gas to the generators. Pressure swing adsorption is the underlying principle behind the working of such concentrating devices [7].

3.2 Ozone Generator

The device employed in the production of ozone is a generator. Ozone gas could be formed mainly by three methods called corona discharge, ultraviolet, and electrolysis.

- Corona discharge method

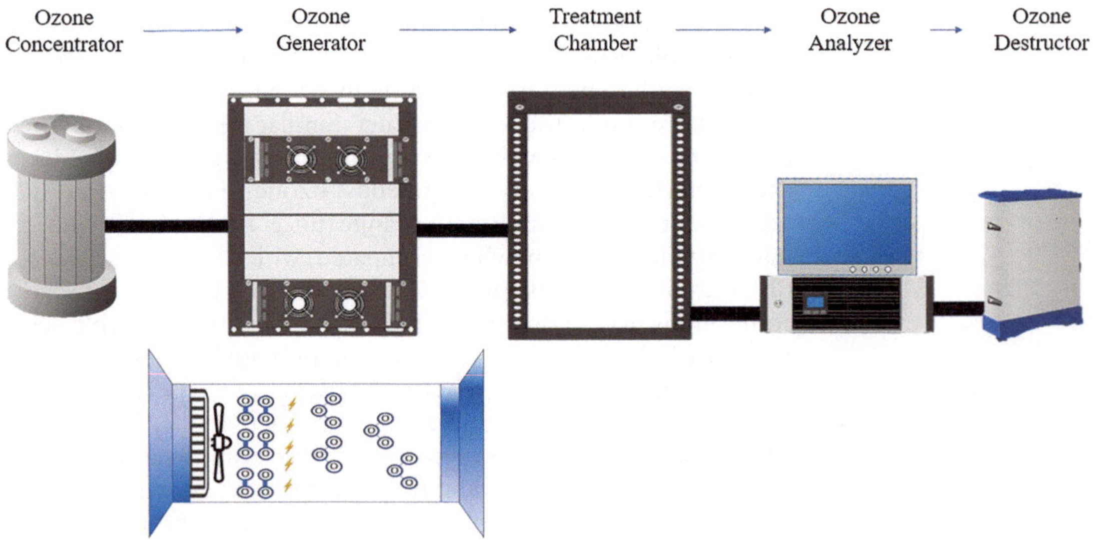

Fig. 1 Ozone generation system

"Silent electric discharge procedure," widely known as corona discharge, is the most commercial method of ozone production [7]. In this technique oxygen-rich feed gas is passed across two electrodes with an applied electric field having an energy >5000 V splitting the gas molecules [49]. Thereof the split-up oxygen radicals will bond with the negatively charged oxygen atoms to produce ozone gas. Moreover, 1–3% ozone is generated with atmospheric air as feed gas, whereas more than 16% yield is formed with pure oxygen feed. The amount of gas generated will depend on certain factors such as the amount of moisture present, gap width, pressure of gas, dielectric properties of electrode material, concentration of oxygen in feed, frequency and voltage of current, and so on [7, 49].

- Ultraviolet method

  UV or physicochemical method of generation is similar to the ozone production in stratosphere, which uses radiation at a wavelength of 185 nm where photodissociation causes oxygen atoms to split. However, for producing ozone industrially, an ultraviolet source with wavelength less than 210 nm could be employed [49]. Additionally, this technique offers an advantage of using ambient gas effectively as the feed gas [7].

- Electrolysis

  Aqueous phosphate solutions subjected to high current density electrolysis will yield ozone as the anodic gas. Nevertheless, electrochemical splitting of water will also result in formation of good amount of oxygen but is not economical due to its higher costs compared to other methods [7].

UV and CD generators are the most commonly used equipment for ozone production. Moreover, the characteristic properties of each food product will determine the ozone concentration it should receive. For instance, grains may be subjected to an ozone concentration of about 10–60 ppm [50–52], fruits and vegetables 5–10 ppm [53–56], and so on. Furthermore, the limitation of ozone in bottled juices is only 0.4 mg/L; nonetheless higher concentrations have not caused any kind of quality deteriorations in the product [57].

3.3 Collector

A collector or treatment chamber is used for exposing the food products to ozone as depicted by Tiwari et al. [25], Pandiselvam et al. [49], and Souza et al. (2018) [55, 56]. The generator outlet will be connected to the chamber and will serve as its inlet, and at the same time, its outlet is given to ozone analyzer and afterwards to the destructor [7]. Additionally, a bubble column apparatus may be employed as the collector for treating liquid foods [58]. Simultaneously, the chamber should be made airtight to prevent any leakages of the ozone, and its design will depend on the characteristic properties of the product being treated.

3.4 Ozone Analyzer

An analyzer is a component used in the ozonation system in order to measure the concentrations of ozone that enter as well as leave the treatment chamber. The concentrations may be analyzed at various desirable points in the chamber. In addition, ozone concentrations in the chamber could be viewed on the analyzer display.

3.5 Ozone Destructor

Ozone gas in excess may be lethal to humans [49], and hence it needs to be destroyed. Thus, an ozone destructor is used when and thereafter the instrument will release clean and ozone-free air into the atmosphere. Furthermore, a safe and nonhazardous by-product is ensured by this equipment [7].

4 Materials and Methods

1. The ozone treatment setup consists of oxygen concentrator/oxygen cylinder, ozone generator, treatment chamber, ozone analyzer, and ozone destructor.
2. All the equipment should be place in line and must be placed at the flat surface.
3. Oxygen concentrator/oxygen cylinder is the inlet to the ozone generator. The ozone concentration depends on the flow rate of oxygen. Hence, the flow rate should be adjusted based on the required ozone concentration. Increase in flow rate decreases the oxygen concentration.
4. The oxygen concentrator must be switched on first and then the ozone generator should be switched on next.
5. The commercially available ozone generator works on the principle of corona discharge. Thus, stabilizer must be connected to the ozone generator in order to supply constant voltage and produce uniform concentration.
6. The ozone gas is generated at high voltage; hence, the generator should be properly covered with suitable material and should not be accessible during operation.
7. Outlet pipe of the ozone generator needs to be connected with the treatment chamber.
8. The ozone treatment chamber should be hermetically (airtight) sealed. The toxic effect of ozone is highlighted in **Notes 1** and **2**.
9. To produce aqueous ozone, diffuser must be connected at the end of the ozone outlet and should be connected in the water source. The selection of proper diffuser size is important to generate microbubbles.

10. Some of the research studies highlighted that gaseous ozone is more effective than aqueous ozone. The selection of the state of ozone depends on the nature of the samples. Gaseous ozone is the best option for the treatment of stored products and dried products (chilies, grapes (resins), etc.). Minimally processed vegetables/fruits and juices can be treated with aqueous ozone.
11. The plenum chamber should be provided at the bottom of the treatment chamber for the treatment of grains.
12. The ozone gas must be released at the bottom of the chamber to ensure the uniform gas distribution.
13. The ozone gas diffusion channel should be provided inside the treatment chamber for the treatment of high bulk density material such as paddy, rice, etc. The treatment details for the management of insects in stored products are presented in **Notes 3** and **4**.
14. The whole surface of fruits and vegetables should be exposed to ozone gas in order to increase the effectiveness of the treatment. Several ozone inlets should be provided in the treatment chamber to reach the desirable concentration within short time. *See* **Notes 5** and **6** for more details.
15. Ozone outlet may be provided at the top of the treatment chamber.
16. Ozone concentration may be calculated by different methods. Ozone analyzer is the common equipment used for the measurement of ozone dose. Some empirical formulas are also used for the measurement of ozone dose. The most common measurement of ozone dose is indicated below

$$\text{Ozone dose}\left(\text{g/m}^3\right) = \frac{\text{Treatment time} \times \text{Ozone Concentration}}{\text{Volume of the treatment Chamber}}$$

$$\text{Dose}(\text{mg/kg min}) = \frac{(\text{Inlet ozone concentration (mg/L)} \times \text{flow rate}(\text{L/min}))}{\text{Product mass(kg)}}$$

17. Ensure that the air filter provided in the ozone analyzer is cleaned frequently.
18. The ozone analyzer outlet pipe should be connected with ozone destructor or water tank to neutralize the excess ozone.
19. Ozone will quickly degrade due to unstable nature; thus, the connectors/connecting pipelines between treatment chamber and ozone analyzer should be short as much as possible. The reaction rate "*k*" must be used to calculate the exact ozone concentration.
20. A high concentration of ozone increases the rate of ozone degradation. Hence, the important ozone process parameters

such as ozone concentration and exposure time should be optimized for each commodity.

21. Monitor the percentage of oxygen and ozone in the treatment chamber to understand the ozone reactivity with the samples.
22. Record the atmospheric temperature and the temperature inside the treatment chamber. The ozone effectiveness also depends on the temperature. Ozone half-life has inverse relationship with temperature. Degradation of ozone is faster at higher temperature, and the low temperature treatment increases the half-life of ozone.
23. In case of aqueous ozone treatment, check the water pH and the presence of organic solvents in the water. Both the parameters are severely affecting the ozone efficacy.
24. Treatment of agro-produces with excess ozone will damage the nutritional profile of the product (*see* **Note 7**).
25. Conveyor speed, material thickness (for solids), ozone dose, temperature, and flow rate should be optimized for the development of continuous type gaseous ozone treatment setup (*see* **Note 8**).
26. Treatment chamber volume, medium (water), mass and density of the samples treated, ozone concentration and contact time; microbubble size, surface area of the sample, and surface nature of the sample (smooth/rough) are important parameters for aqueous ozone treatment.
27. Record the mass, density, porosity, and temperature of the sample before the operation.
28. Start the treatment process, monitor the flow rate of oxygen and ozone and concentration of ozone, and ensure that constant voltage supplies the ozone generator during the operation (*see* **Note 9**).
29. Perform the visual observation of the ozone-treated samples to check the undesirable changes on the surface or color (*see* **Note 5**).
30. Ozone is having a high oxidation potential; hence, the pipes/connectors used in the ozone treatment system should not be rubber or plastic (*see* **Note 10**).
31. Food-grade stainless steel is the suitable material for the design of treatment chamber.
32. As a precautionary measure, ozone detector must be attached in the ozone setup room to detect the ozone leakage.

5 Notes

1. Ozone gas is toxic to human beings. Hence, it is very important that the human beings should not be exposed to ozone during experiment.
2. Also, the treatment chamber should not be opened immediately after the ozone treatment. Ozone gas will be autodegraded after certain time depending on the concentration. Otherwise, ensure proper degassing arrangement after treatment. Ensure that the ozone setup is placed in the proper ventilated place.
3. The known number of insects should be placed at different depth of the treatment chamber to understand the ozone efficacy. Ensure that the control or untreated samples should be handled in similar manner for results comparison.
4. Treatment for stored product insects required high ozone dose and exposure time than microorganisms/fungi.
5. The surface color change in the ozone-treated produces was observed in fresh-agro produces. Treatment of the produces at high ozone concentration may also change the flavor of the product.
6. The treatment of whole products is preferred for ozone treatment than cut/sliced fruits/vegetables to avoid oxidative tissue damage at high ozone dose.
7. The high moisture foods will react with ozone more rapidly than medium and low moisture foods. Hence, the lower ozone dose may be applied for high moisture foods.
8. The material thickness (bed thickness) and moisture content of the samples have positive relation with ozone destruction. The half-life of the ozone is higher at low moisture samples than high moisture samples.
9. Immediately after switching on the equipment, check any ozone gas leaks from the connectors, valves, or treatment chamber. Switch off the equipment, if there is any gas leakage.
10. Clean the treatment chamber with water or standard solution and dry it before next use.

References

1. Mahapatra AK, Muthukumarappan K, Julson JL (2005) Applications of ozone, bacteriocins and irradiation in food processing: a review. Crit Rev Food Sci Nutr 45:447–461
2. Güzel-Seydim ZB, Greene AK, Seydim AC (2004) Use of ozone in the food industry. LWT Food Sci Technol 37:453–460
3. Hirahara Y, Iwata K, Nakamuro K (2019) Effect of citric acid on prolonging the half-life of dissolved ozone in water. Food Saf 7:90–94
4. Donnell CO, Cullen PJ (2012) Ozone in food processing. John Wiley & Sons, Chichester
5. Miller FA, Silva CLM, Brandão TRS (2013) A review on ozone-based treatments for fruit and

vegetables preservation. Food Eng Rev 5: 77–106

6. Rice RG, Graham DM (2001) U.S. FDA regulatory approval of ozone as an antimicrobial agent – what is allowed and what needs to be understood. Ozone News 29:22–31
7. Prabha V, Barma RD, Singh R et al (2015) Ozone technology in food processing: a review. Trends Biosci 8:4031–4047
8. Zhu F (2018) Effect of ozone treatment on the quality of grain products. Food Chem 264: 358–366
9. Gonçalves AA, Kechinski CP (2011) Ozone technology in the food industry. In: Food engineering. Nova Science, New York, NY, pp 85–146
10. Garud SR, Priyanka BS, Rastogi NK et al (2018) Efficacy of ozone and lactic acid as nonthermal hurdles for preservation of sugarcane juice. Ozone Sci Eng 40:198–208
11. Dubuis ME, Dumont-Leblond N, Laliberté C et al (2020) Ozone efficacy for the control of airborne viruses: bacteriophage and norovirus models. PLoS One 15:1–19
12. Klein H-P (1988) Ozone in water treatment processes. In: Process technologies for water treatment. Plenum, New York, NY, pp 145–156
13. Brodowska AJ, Nowak A, Śmigielski K (2018) Ozone in the food industry: principles of ozone treatment, mechanisms of action, and applications: an overview. Crit Rev Food Sci Nutr 58:2176
14. Kroschwitz JI, Seidel A (2005) Kirk-Othmer encyclopedia of chemical technology. Wiley-Interscience, New York, NY
15. Triandi Tjahjanto R, Galuh RD, Wardhani S (2012) Ozone determination: a comparison of quantitative analysis methods. J Pure Appl Chem Res 1:18–25
16. Sarron E, Gadonna-Widehem P, Aussenac T (2021) Ozone treatments for preserving fresh vegetables quality: a critical review. Foods 10: 605
17. Bader H, Hoigné J (1981) Determination of ozone in water by the indigo method. Water Res 15:449–456
18. Chen C, Liu C, Jiang A et al (2020) Effects of ozonated water on microbial growth, quality retention and pesticide residue removal of fresh-cut onions. Ozone Sci Eng 42:399
19. Ziyaina M, Rasco B (2021) Inactivation of microbes by ozone in the food industry: a review. Afr J Food Sci 15:113
20. OSHA (n.d.) Occupational Safety and Health Administration's (OSHA) regulations for ozone. Occupational Safety and Health Administration. https://www.osha.gov/laws-regs/standardinterpretations/1994-09-29-0
21. Ahmad M, Farooq S (1985) Influence of bubble sizes on ozone solubility utilization and disinfection. Water Sci Technol 17:1081
22. Patil S, Cullen PJ, Kelly B et al (2009) Extrinsic control parameters for ozone inactivation of Escherichia coli using a bubble column. J Appl Microbiol 107:830–837
23. Kaess G, Weidemann JF (1973) Effects of ultraviolet irradiation on the growth of microorganisms on chilled beef slices. Int J Food Sci Technol 8:59
24. Castillo A, McKenzie KS, Lucia LM et al (2003) Ozone treatment for reduction of Escherichia coli O157:H7 and Salmonella serotype Typhimurium on beef carcass surfaces. J Food Prot 66:775
25. Tiwari BK, Muthukumarappan K, O'Donnell CP et al (2008) Modelling colour degradation of orange juice by ozone treatment using response surface methodology. J Food Eng 88:553–560
26. Galgano F, Caruso MC, Condelli N et al (2015) Application of ozone in fresh-cut iceberg lettuce refrigeration. Adv Hortic Sci 29: 61
27. Marino M, Maifreni M, Baggio A et al (2018) Inactivation of foodborne bacteria biofilms by aqueous and gaseous ozone. Front Microbiol 9:2024
28. Ishizaki K, Shinriki N, Matsuyama H (1986) Inactivation of Bacillus spores by gaseous ozone. J Appl Bacteriol 60:67
29. Rice RG, Robson CM, Miller GW et al (1981) Uses of ozone in drinking water treatment. J Am Water Works Assoc 73:44
30. Steenstrup LD, Floros JD (2004) Inactivation of E. coli O157:H7 in apple cider by ozone at various temperatures and concentrations. J Food Process Preserv 28:103
31. Wu J, Doan H, Cuenca MA (2006) Investigation of gaseous ozone as an anti-fungal fumigant for stored wheat. J Chem Technol Biotechnol 81:1288
32. Coll Cárdenas F, Andrés S, Giannuzzi L et al (2011) Antimicrobial action and effects on beef quality attributes of a gaseous ozone treatment at refrigeration temperatures. Food Control 22:1442
33. Cho Y, Muhlisin M, Choi JH et al (2014) Effect of gaseous ozone exposure on the bacteria counts and oxidative properties of ground hanwoo beef at refrigeration temperature. Korean J Food Sci Anim Resour 34:525

34. Liu C, Chen C, Jiang A et al (2021) Effects of aqueous ozone treatment on microbial growth, quality, and pesticide residue of fresh-cut cabbage. Food Sci Nutr 9:52–61
35. Cantalejo MJ, Zouaghi F, Pérez-Arnedo I (2016) Combined effects of ozone and freeze-drying on the shelf-life of Broiler chicken meat. LWT Food Sci Technol 68:400
36. Habibi Najafi MB, Haddad Khodaparast MH (2009) Efficacy of ozone to reduce microbial populations in date fruits. Food Control 20:27
37. Brodowska AJ, Smigielski K, Nowak A et al (2015) The impact of ozone treatment in dynamic bed parameters on changes in biologically active substances of juniper berries. PLoS One 10:e0144855
38. Sarig P, Zahavi T, Zutkhi Y et al (1996) Ozone for control of post-harvest decay of table grapes caused by Rhizopus stolonifer. Physiol Mol Plant Pathol 48:403
39. Sotelo JL, Beltran FJ, Benitez FJ et al (1987) Ozone decomposition in water: kinetic study. Ind Eng Chem Res 26:39
40. Chen CL, Chang HM, Gratzl JS et al (1984) Studies an ozone bleaching. I. The effect of PH, temperature, buffer systems and heavy metal-ions on stability of ozone in aqueous solution. J Wood Chem Technol 4:367
41. Patil S, Valdramidis VP, Cullen PJ et al (2010) Inactivation of Escherichia coli by ozone treatment of apple juice at different pH levels. Food Microbiol 27:835
42. Ferreira WFS, Alencar ER, Blum LEB et al (2020) Ozonation of Brazil nuts in aqueous media at different pH levels: ozone decomposition, aspergillus flavus inactivation, and effects on nut color and crude oil lipid profile. Ozone Sci Eng 43:351
43. Britton HC, Draper M, Talmadge JE (2020) Antimicrobial efficacy of aqueous ozone in combination with short chain fatty acid buffers. Infect Prev Pract 2:10032
44. Khadre, M. A., & Yousef, A. E. (2001). Sporicidal action of ozone and hydrogen peroxide: a comparative study. International journal of food microbiology, 71(2-3), 131–138.
45. Guzel-Seydim, Z. B., Greene, A. K., & Seydim, A. C. (2004). Use of ozone in the food industry. LWT-Food Science and Technology, 37 (4), 453–460.
46. Restaino, L., Frampton, E.W., Hemphill, J. B., & Palnikar, P. (1995). Efficacy of ozonated water against various food-related microorganisms. Applied and environmental microbiology, 61(9), 3471–3475.
47. Williams, E. J., Fehsenfeld, F. C., Jobson, B. T., Kuster, W. C., Goldan, P. D., Stutz, J., & McClenny, W. A. (2006). Comparison of ultraviolet absorbance, chemiluminescence, and DOAS instruments for ambient ozone monitoring. Environmental science & technology, 40(18), 5755–5762.
48. Cullen, P. J., Valdramidis, V. P., Tiwari, B. K., Patil, S., Bourke, P., & O'donnell, C. P. (2010). Ozone processing for food preservation: an overview on fruit juice treatments. Ozone: Science & Engineering, 32(3), 166–179.
49. Pandiselvam R, Sunoj S, Manikantan MR et al (2017) Application and kinetics of ozone in food preservation. Ozone Sci Eng 39:115–126
50. Kells SA, Mason LJ, Maier DE et al (2001) Efficacy and fumigation characteristics of ozone in stored maize. J Stored Prod Res 37: 371–382
51. Mendez F, Maier DE, Mason LJ et al (2002) Penetration of ozone into columns of stored grains and effects on chemical composition and processing performance. J Stored Prod Res 39: 33–44
52. Trombete F, Minguita A, Porto Y et al (2016) Chemical, technological, and sensory properties of wheat grains (Triticum aestivum L) as affected by gaseous ozonation. Int J Food Prop 19:2739–2749
53. Palou L, Crisosto CH, Smilanick JL et al (2002) Effects of continuous 0.3 ppm ozone exposure on decay development and physiological responses of peaches and table grapes in cold storage. Postharvest Biol Technol 24: 39–48
54. Liu C, Ma T, Hu W et al (2016) Effects of aqueous ozone treatments on microbial load reduction and shelf life extension of fresh-cut apple. Int J Food Sci Technol 51:1099–1109
55. Souza LP, Rita L, Faroni DA et al (2018) Effects of ozone treatment on postharvest carrot quality. LWT Food Sci Technol 90:53–60
56. Souza LP, Faroni LRDA, Heleno FF et al (2018) Ozone treatment for pesticide removal from carrots: optimization by response surface methodology. Food Chem 243:435–441
57. Pandiselvam R, Subhashini S, Priya EPB et al (2018) Ozone based food preservation: a promising green technology for enhanced food safety. Ozone Sci Eng 41:17–34
58. Panigrahi C, Mishra HN (2020) Effect of ozonation parameters on nutritional and microbiological quality of sugarcane juice. J Food Process Eng 43:e13542

Chapter 5

Application of Electrolyzed Water as Disinfecting Agent in Table Egg to Decrease the Incidence of Foodborne Pathogens

Juan C. Ramírez-Orejel and José A. Cano-Buendía

Abstract

Egg is one of the most commonly consumed foods worldwide due to its nutritional quality; additionally, it has technological properties that can be implemented in the food industry. To ensure hygiene, eggs are treated (washed). Pasteurization is also another treatment option; however, heat treatment affects egg physical properties. An alternative treatment is the use of neutral electrolyzed water as a decontaminating agent, which has already been tested in different foods, including table eggs. Neutral electrolyzed water has generated favorable results in terms of bacterial burden reduction, in the improvement of physicochemical and functional properties of food.

In this chapter, the procedures for monitoring the effect of washing eggs with neutral electrolyzed water are described. The washing procedure described focuses on laboratory testing; however, large-scale industrial procedures can be adapted for high-throughput performance.

Key words Neutral electrolyzed water, Foodborne pathogens, Table egg, Egg disinfection

1 Introduction

Currently, the food industry and health sector are facing increases in the incidence of foodborne illnesses caused by pathogenic microorganisms. This type of contamination can occur in raw, pasteurized, or ready-to-eat products, such as table egg.

Egg consumption is growing [1] given that it is an inexpensive and highly nutritious food product. Egg provides important elements such as essential amino acids [2], and it has various technological properties (like a foaming agent or because egg yolk can act as emulsifier) which are applicable in the food industry.

Eggs can be contaminated with pathogenic bacteria like *Escherichia coli*, *Salmonella* sp., and *Listeria* sp. among others, which are the causative agents of infectious diseases affecting the egg industry [3].

Mohsen Gavahian (ed.), *Emerging Food Processing Technologies*,
Methods and Protocols in Food Science, https://doi.org/10.1007/978-1-0716-2136-3_5,

Egg contamination can be caused by diverse factors including hen health and breeding conditions that influence the overall load and diversity of bacterial species. Furthermore, cracked eggshells may also increase the risk of contamination; therefore controlling factors influencing the presence of cracked eggshell is essential to prevent the spread of microorganisms [4].

Heat treatment is sufficient to eradicate potential pathogens in food; however, *Escherichia coli* O157:H7 has a thermal death point of 70 °C; thus this serotype may resist the effects of pasteurization.

To eliminate potential pathogens in eggs, the poultry industry has opted for the use of whole egg disinfectants, which includes sodium hypochlorite, sodium hydroxide, formaldehyde, and citric acid, among others; nevertheless, such disinfectants can be corrosive and toxic; additionally, heat treatment modifies the physical properties of table eggs.

The use of neutral electrolyzed water (NEW) is an attractive alternative due to its optimal characteristics: antibacterial activity; easy handling; environmental friendly [5]; and null effects on egg cuticle [6].

The use of NEW as a disinfectant solution on eggs has several industrial applications: ensure hygiene for consumption; increase shelf life [6, 7]; and preparation of viable eggs for incubation or hatching chambers.

The production of NEW consists of a mixture of tap water and a low concentration sodium chloride (NaCl) solution, after which, the mixture is electrolyzed (Fig. 1). The electrolysis chamber is separated by a membrane or diaphragm. NaCl molecules are then dissociated into ionic forms, namely, Na^+ and Cl^- ions. The Na^+ ions are attracted towards the cathode and the Cl^- ions to the anode. Similarly, water undergoes two types of dissociation processes depending on the medium. Hydroxyl ions and molecular hydrogen will be formed where the cathode is located; the hydroxyl ions will react with Na^+ to generate sodium hydroxide, thus producing alkaline electrolyzed water with a pH of 11–13 [8]. On the anode side, water will dissociate into protons and molecular oxygen, and the hydroxyl and chloride ions will be absorbed by the membrane towards the electrode side; at the same time, the hydroxyl and chloride ions release an electron to form radicals that when combined will generate hypochlorous acid, ultimately generating acidic electrolyzed water with pH of 2.5–4.5 [9]. The production of NEW with a pH of approximately 6.4–7.5, and an oxidation reduction potential of 800–900 mV, is carried out by a controlled mixture of the abovementioned acidic and alkaline electrolyzed waters.

This document encompasses the procedures to consider when evaluating NEW prior to its application and when monitoring for

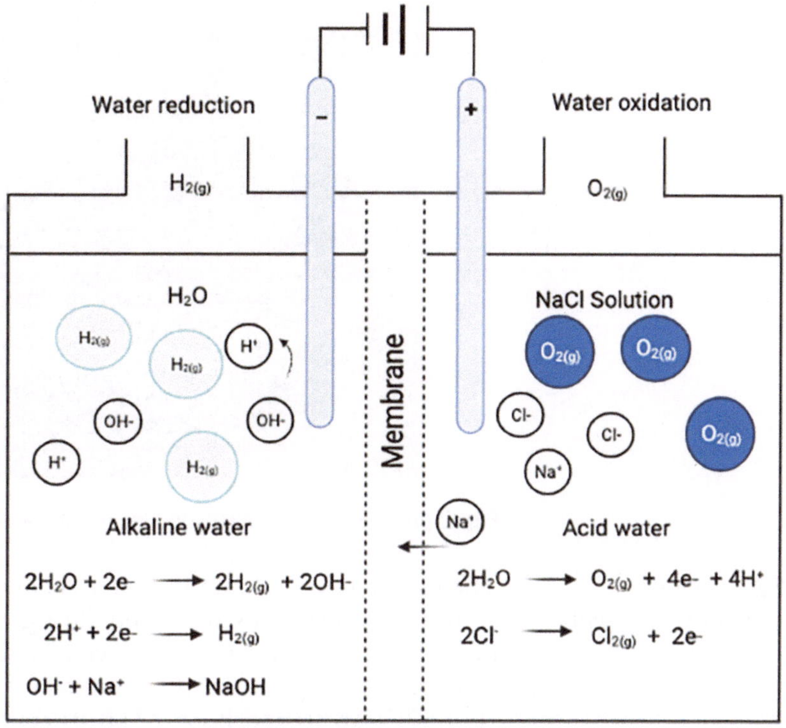

Fig. 1 Electrolyzed water production

remaining bacteria after treatment. The procedures can be modified when electrolyzed water is to be implemented in other food types.

2 Materials and Methods

NEW can be diluted with tap water; thus water quality should be evaluated.

NEW can be purchased or generated on-site using equipment from different suppliers. Diluted and non-diluted electrolyzed water can be stored at room temperature. During preparation and storage, follow all applicable waste disposal regulations. The use of NEW offers the advantage that electrolyzed water reacts with organic material generating water and sodium chloride as final products.

2.1 Oxidation Reduction Potential and pH Measurements

1. Pour 50 mL in a 250 mL flask. If possible, use a magnetic bar to stir.
2. Use a pH/ORP/temperature combo tester (HANNA, HI98121) or a similar tester (Fig. 2) to obtain pH and ORP values. Neutral electrolyzed water has a pH range between 6.5 and 7.8 and ORP values higher than 750 [10].

Fig. 2 pH/ORP meter to evaluate electrolyzed water

2.2 Chlorine Measurement (ppm)

The use of a portable photometer (Hanna Instruments HI96771) is more advantageous when measuring active chlorine concentration, as its sensitivity ranges between 0 and 500 ppm. The following is a brief description of the manufacturer's protocol: reagents A (HI95771AB-0) and B (95771B-0) are added to the sample photocell, and then the meter is calibrated for 1 min; afterwards, an electrolyzed water sample is placed in another photocell, and reagents A and B are added; then chlorine concentration can be read.

2.3 Egg Wash

The egg wash can be performed manually or with specialized industrial washing systems. In this document the spray bottle procedure is described (*see* **Note 1**). If eggs are contaminated with a specific pathogenic bacterium [11], the procedure will need to be performed in a biosafety cabinet; otherwise it can be performed in any designated clean working area. The procedure can be carried out on embryonated eggs also (*see* **Note 2**).

1. Eggs are placed vertically in plastic egg trays. Trays can be set over a container to collect wash solution.
2. Neutral electrolyzed water concentration can be adjusted with tap water when filling plastic spray bottles. We recommend using 15 mL/10–15 eggs.
3. Apply half of the treatment, allowing solution to cover entire egg surface.
4. Use sterile gloves and turn all treated eggs vertically.
5. Apply remaining half of treatment solution to eggs.

2.4 Bacterial Count

The bacterial count is a crucial parameter when evaluating egg wash effectiveness. Sterile gloves should be worn at all times when handling eggs. It is important to utilize tryptic soy agar (TSA) Petri dishes for this evaluation (*see* **Note 3**), a microbiology incubator is also required, and always work in a clean area with a Bunsen burner to assure sterile conditions (*see* **Note 4**).

1. Introduce individual eggs into sterile plastic bags (Nasco Whirl-Pak, B01065WA).
2. Add 30 mL of sterile saline solution (*see* **Note 5**) to each plastic bag and gently massage for 1 min.
3. Perform serial decimal dilutions of washed saline solution with sterile PBS and plate 100 μL on TSA Petri dishes. The first dilution needs to be performed with peptone water (*see* **Note 6**); following dilutions can be performed with saline solution (*see* **Note 7**).
4. Spread samples using glass beads or L-shaped disposable spreaders under sterile conditions (*see* **Note 8**).
5. Incubate Petri dishes overnight at 37 °C.
6. Count bacterial colonies when numbers are between 20 and 200 CFU/Petri dish. Bacterial counts need to be multiplied by dilution factor to identify bacterial load per egg.

2.5 Haugh Units

Haugh units (HU) are related to the quality of table eggs. HU can be obtained after treatment and/or storage.

1. Crack egg and place content on a flat surface.
2. Use a Haugh's micrometer (Baxlo) (Fig. 3) and take five different readings of the dense albumen.

Fig. 3 Measuring Haugh units from dense albumen

3. Use the following formula:

$$= 100 * \log(h{-}^{0.37} + 7.6)$$

where

h = height of the dense albumen (mm).

w = egg weight (g).

3 Notes

1. High-density polypropylene (HDPP) spray bottles are recommended when performing manual egg wash procedure.
2. Temperature is a critical factor to consider during the sanitization process of embryonated eggs, to avoid any detrimental effects on hatching percentage. NEW can be applied using thermal fogging devices.
3. TSA is a general growth medium. TSA components are pancreatic digest of casein (15 g), peptic digest of soybean meal (5 g), sodium chloride (5 g), and agar (15 g). These ingredients are added to 1 L of deionized water. Sterilize at 121 °C for 15 min.
4. Bacteriological procedures need to be performed using a Bunsen burner in a clean and closed working area.
5. Saline solution components: NaCl (9 g) in 1000 mL of deionized water. Sterilize at 121 °C for 15 min.
6. Peptone water components: Peptone Bacteriological (1.0 g) in 1000 mL of deionized water. Sterilize at 121 °C for 15 min.
7. Vortex when performing serial dilutions when performing each dilution.
8. Electrolyzed water activity is inactivated in the presence of peptone. Therefore, paper test discs are not recommended when electrolyzed water is evaluated.

References

1. McDougal T (2020) Global egg production continues to rise. In: Poult. World. https://www.poultryworld.net/Eggs/Articles/2020/6/Global-egg-production-continues-to-rise-604164E/. Accessed 15 Feb 2020
2. Kuang H, Yang F, Zhang Y et al (2018) The impact of egg nutrient composition and its consumption on cholesterol homeostasis. Cholesterol 2018:6303810. https://doi.org/10.1155/2018/6303810
3. Turtoi M, Borda D (2014) Decontamination of egg shells using ultraviolet light treatment. Worlds Poult Sci J 70:265–278. https://doi.org/10.1017/S0043933914000282
4. Wales A, Breslin M, Carter B et al (2007) A longitudinal study of environmental salmonella contamination in caged and free-range layer

flocks. Avian Pathol 36:187–197. https://doi.org/10.1080/03079450701338755

5. Orejel JCR, CanoBuendía JA (2020) Applications of electrolyzed water as a sanitizer in the food and animal-by products industry. PRO 8: 534. https://doi.org/10.3390/pr8050534
6. Medina-gudiño J, Rivera-garcia A, Santos-ferro L et al (2020) International journal of food microbiology analysis of neutral electrolyzed water anti-bacterial activity on contaminated eggshells with Salmonella enterica or Escherichia coli. Int J Food Microbiol 320: 108538. https://doi.org/10.1016/j.ijfoodmicro.2020.108538
7. Rivera-Garcia A, Santos-Ferro L, Ramirez-Orejel JC et al (2019) The effect of neutral electrolyzed water as a disinfectant of eggshells artificially contaminated with Listeria monocytogenes. Food Sci Nutr fsn3.1053. https://doi.org/10.1002/fsn3.1053
8. Fasenko GM, O'Dea Christopher EE, McMullen LM (2009) Spraying hatching eggs with electrolyzed oxidizing water reduces eggshell microbial load without compromising broiler production parameters. Poult Sci 88: 1121–1127. https://doi.org/10.3382/ps.2008-00359
9. Ozer NP, Demirci A (2006) Electrolyzed oxidizing water treatment for decontamination of raw salmon inoculated with Escherichia coli O157:H7 and listeria monocytogenes Scott A and response surface modeling. J Food Eng 72: 234–241. https://doi.org/10.1016/j.jfoodeng.2004.11.038
10. Yan P, Daliri EBM, Oh DH (2021) New clinical applications of electrolyzed water: a review. Microorganisms 9:1–21. https://doi.org/10.3390/microorganisms9010136
11. Bialka K, Demirci A, Knabel S et al (2004) Efficacy of electrolyzed oxidizing water for the microbial safety and quality of eggs. Poult Sci 83:2071–2078

Chapter 6

Combination of Electrolyzed Water and Other Measures for Food Decontamination: Methods and Procedures

S. M. E. Rahman, H. M. Murshed, and S. M. A. Islam

Abstract

Food safety is a burning issue in the present world. Safe sanitizers are obligatory for maintaining quality of food and increasing the shelf life of fresh produce and other agricultural products. Food industries have been using electrolyzed water (EW) as a unique sanitizer for the past two decades which has excellent results to reduce the microbial count. Hurdle technology, e.g., combination of EW with ultrasonication, short-term heat treatment, organic acids, and salts, found to have more effective results in reducing microorganisms which overcame the little shortcomings with EW like corrosiveness and maintained organoleptic qualities. In this chapter, we are going to discuss the production of EW and its combination with ultrasonication, short-term heat treatment, organic acids, and salts to produce a novel sanitizer.

Key words Electrolyzed water (EW), Combination, Hurdle technology, Safety, Sanitizer

1 Introduction

Disease outbreak from food and related products has arrived as an acute health problem throughout the world. The Centers for Disease Control and Prevention (CDC) estimates that in human there are 31 varieties of different foodborne pathogens causing various infection in the USA alone which in turn causes illness of 48 million people and 841 outbreaks of foodborne disease and causes death of 3000 people annually [1]. Hence, production of operative disinfectants for reducing pathogens is a key step for the HACCP (hazard analysis and critical control point) systems in food manufacturing [2]. The agricultural and food production industry has used many decontamination strategies across different food production chain. Nevertheless, many of the techniques of decontamination processes have remarkable drawbacks such as costly program, chemical residue, little efficacy, and hostile outcomes for superiority of different food-related products [3]. Therefore, a sanitizer should be free of hazardous chemical compounds, and

Mohsen Gavahian (ed.), *Emerging Food Processing Technologies*,
Methods and Protocols in Food Science, https://doi.org/10.1007/978-1-0716-2136-3_6,

safe sanitizers in food industry are indispensable to ensure quality and safety of food products [4, 5].

In recent years, electrolyzed water (EW) has become the use as cleaner (EW and NaOH mixture) as well as sanitizer (EW and HOCl mixture). The extensive use of EW in food industry is because of its popularity. The core reason for its admiration is the ease of production and implementation. Diluted NaCl (in some cases $MgCl_2$ or KCl) in electrolyzed chamber is used to prepare EW. Ordinary water is mostly used for EW production without inclusion of harmful substances except NaCl [6]. However in Russia the concept of EW was found from long before, and it had been extensively exercised in Japan since 1980 in different medical institutions mainly for water decontamination, regeneration, and disinfection [7–9]. Utilization of EW started popularly in different food industries in the past two decades for hygienic control owing to its active anti-pathogenic as well as antimicrobial action and its low operating cost [10, 11].

EW has been successfully used in several experiments which found effective results on reducing microbial count such as on food processing surfaces and non-food surfaces [9, 12–14], food handling gloves [15], cutting boards [16], shrimp [17], fish [18], beef [18, 19], pork [20, 21], poultry carcasses [18, 20], fruits [12, 22], and vegetables [23, 24]. Slightly acidic electrolyzed water (SAEW) and alkaline electrolyzed water (AlEW) have been extensively practiced as stated above. Likewise, many different hurdle practices have been broadly used in the food industry in disinfecting process. Hurdle technology is a term where two or more preservation or disinfecting techniques are combined in order to minimize the sensory loss of food [25]. Keeping this concept in mind, scientists started combination of EW with ultrasonication, organic acid, thermal process, and different salts in disinfecting process. These combinations of studies found to have more effective results in the reduction of microbes such as alkaline electrolyzed water in mild heat and citric acid combined treatment [26]. Ultrasonication enhanced low concentration electrolyzed water [27], treatment from EW at low concentration and short heat period [28], and combined treatment of calcium lactate and EW at low concentration [29].

Considering the above studies, it can be said that combined action of EW with ultrasonication, short-time heat treatment, organic acids, and salts can be suitable hurdle technologies in decreasing foodborne pathogens in food industry and thus can increase the shelf life of the agricultural products with better storage quality. Therefore, the methods of combination techniques of EW with ultrasonication, short-time heat treatment, organic acids, and salts and sanitizer production techniques are discussed in this chapter.

2 Materials

2.1 Electrolyzed Water Production

2.1.1 Alkaline and Acidic Electrolyzed Water (AIEW and AEW)

The device used to produce alkaline and acidic electrolyzed water was electrolysis device, ROX-10WB3, Hoshizaki Electric Company, Aichi, Japan [28].

2.1.2 Slightly Acidic Electrolyzed Water (SAEW)

To produce SAEW, electrolysis device (without membrane, single chamber), BC-360, Cosmic Round Korea Co. Ltd., Seongnam, Korea [27], was used.

2.1.3 Neutral Electrolyzed Water (NEW)

For production of NEW, Eurostel R EZ-90 Unit (Ecanet, Palamos, Girona, Catalonia, Spain) [30] was used.

2.2 Organic Acids

The organic acids used were propionic acid, malic acid, levulinic acid, acetic acid [31], tartaric acid, lactic acid [31], citric acid [32], fumaric acid [31], etc.

2.3 Ultrasonicator

The ultrasonicator machine is a benchtop ultrasonic cleaner (JAC-4020, KODO Technical Research Co., Ltd., Hwaseong, Gyeonggi-do, South Korea) [27].

2.4 Salts

The salts used were calcium lactate (CaL) [29], potassium sorbate [33], sodium lactate [33], sodium acetate [33], sodium citrate [33], sodium chloride [33], etc.

3 Methods

3.1 Electrolyzed Water (EW) Production

Generally, electrolysis chamber is used which contains NaCl solution (different concentration as per required). A diaphragm (membrane or septum) divides the chamber into cathode and anode [9]. At EW generator, the current and voltage value is set at 8–10 A and 9–10 V [2], respectively, between the electrodes. During electrolysis process, NaCl breaks into ions for positive (Na^+) and negative (Cl^-). At the same time, in solution hydrogen (H^+) ions and hydroxide (OH^-) ions are also produced. At anode, the negative ions (OH^- and Cl^-) accumulate as well as release electrons, and hypochlorous acid (HOCl), hypochlorite ion (OCl^-), oxygen gas (O_2), hydrochloric acid (HCl), and chlorine gas (Cl_2) are also produced. Simultaneously at cathode, positive (H^+ and Na^+) ions accumulate and get electrons, and hydrogen gas (H_2) and sodium hydroxide (NaOH) are produced [3, 9]. At a time, two types of EW are generated (Fig. 1). A solution which is acidic in nature is found at anode which have ORP (oxidation reduction potential) of >1100 mV, pH of 2–3, and ACC (available chlorine concentration) at 10–90 ppm [2]. This acidic solution is AEW (acidic electrolyzed

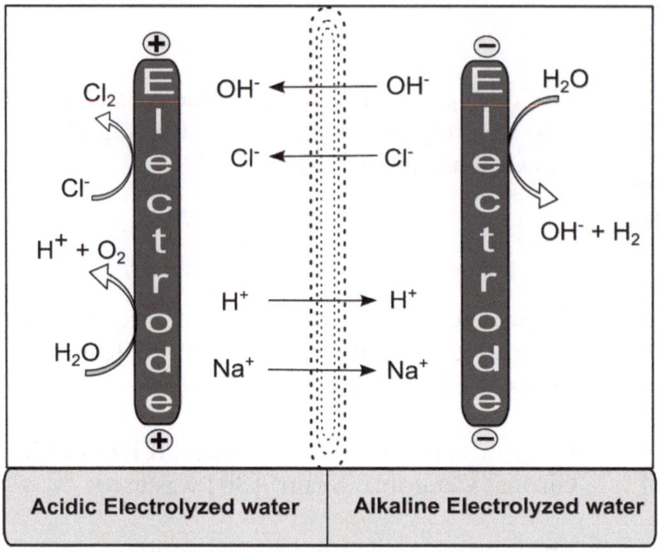

Fig. 1 Schematic diagram of acidic and alkaline electrolyzed water generation

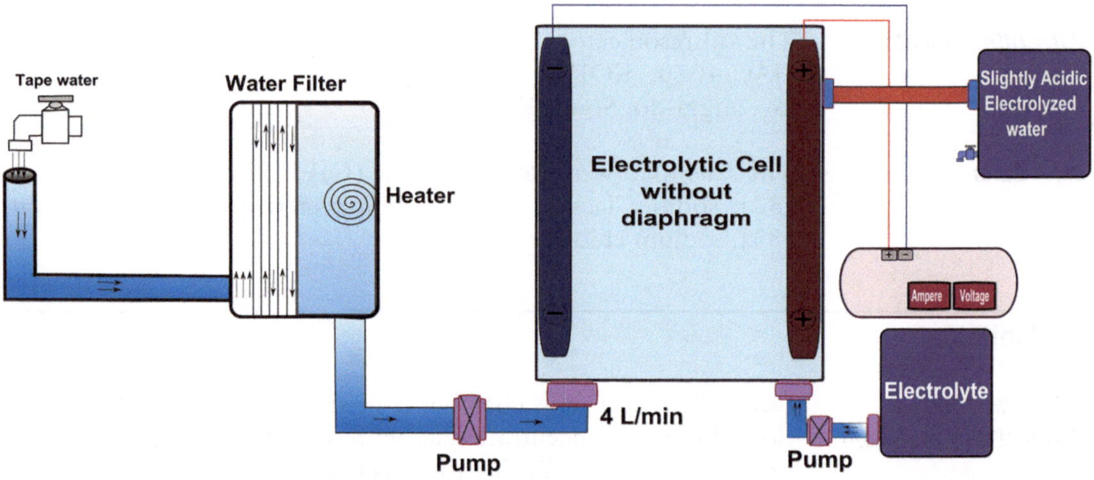

Fig. 2 Schematic diagram of slightly acidic electrolyzed water generation

water) or EOW (electrolyzed oxidizing water). Similarly, from cathode, a basic solution having pH 10–13 and ORP value of −800 to −900 mV is formed [2]. This basic solution is referred as alkaline electrolyzed water (AlEW) or electrolyzed reducing water (ERW). From the study of [3, 34], NEW (neutral electrolyzed water) is generated having pH of 7–8 and ORP of 750–900 mV. Slightly acidic electrolyzed water (SAEW) is also produced which have pH of 5–6.5 and ORP of approximately 850 mV [35]. A single cell unit (without diaphragm) using NaCl or HCl solution or mixing anodic solution with OH^- ions is used to prepare NEW [9]. To prepare SAEW, electrolysis of HCl singly or NaCl and HCl combination is done in a single cell unit (without membrane) (Fig. 2) [36].

3.2 Preparation of Organic Acid Solution and Combination with EW

Different types of organic acid solutions can be prepared in the same way. For example, crystalline citric acid is dissolved in sterile distilled water (w/v) at specific amount to get desired concentration, i.e., 1% CA solution [25]; 0.5% fumaric acid solution [37]; and 0.125%, 0.25%, and 0.5% of FA solutions (w/v) [31]. Combination can be done either EW is mixed with the organic acid solution or crystalline organic acid can be dissolved in EW directly [32]. The combination of organic acid solution and EW sanitizing program is conducted by dipping the sample (like carrot) [8] for different or certain duration.

3.3 Combination of Ultrasonication and EW

A benchtop ultrasonic cleaner is usually used for ultrasonication. Generally, in ultrasonication normal water is used, but here we will use EW in lieu of water as we want the ultrasonication of EW. Ultrasonication with EW will be done at 40 kHz of fixed frequency and acoustic energy density (AED) 100, 200, and 400 W/L at room temperature [27]. The duration (1 min [27]; 10 min) [12], amount of EW (10 L) [12], and size of ultrasonic chamber will depend on the type of sample and intensity of ultrasonication. Different sample(s) are usually sanitized by immersing into the ultrasonic chamber containing EW [12].

3.4 Combination of EW and Thermal Process

To combine the thermal and EW, EW will be taken in a water bath, and temperature will be controlled immediately at 60, 70, and 80 °C [28] or desired temperature (*see* **Note 1**). Amount of EW will vary with the sample. Samples will be immersed into the water bath containing EW where temperature can be controlled at desired level [28].

3.5 Combination of Salt and EW

Salt (calcium lactate (CaL), potassium sorbate, sodium chloride) solutions of different concentrations (3% CaL [29], 5% potassium sorbate, 5% sodium chloride [33]) will be prepared by distilled water or mixed with EW to get the combined effect of treatments. To sanitize the sample, it will be dipped into the combined mixture of salt solution and EW (*see* **Notes 2** and **3**) [29].

4 Notes

1. Long-term heat treatment might affect the texture and taste of some food. Also, it may cause a severely damaging effect on the sensory and beneficial properties of some fresh produce [38, 39].
2. Acidic EW is corrosive in nature. Chlorine must be handled with attention as high concentrations may be corrosive. It can irritate the skin or can cause respiratory tract irritation. Compounds such as trihalomethanes or chlorohydroxyfuranones

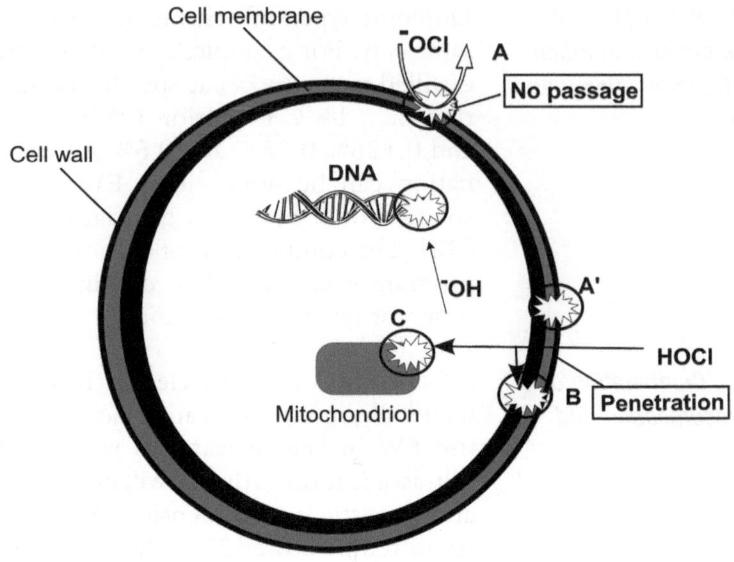

Fig. 3 Germicidal mechanism of electrolyzed water

might be produced from the reaction of HOCl and OCl^- with the organic substances. These compounds might be carcinogenic to humans as they are cytotoxic and genotoxic in nature [40].

3. Bactericidal activity of EW (Fig. 3) is influenced by the presence of organic matter due to its reaction particularly with protein [41], water hardness, and pollutants in the product.

5 Conclusion

EW shared with ultrasonication, short-term heat treatment, salts, and organic acids provides strong bactericidal, fungicidal, and virucidal possessions in many areas such as health sector, on the surface of fresh produce, livestock products and utensils, etc. Nevertheless, acidic EW is corrosive in nature and affects the sensory characteristics, and long-term heat affects the taste of some foods, which limit their uses. Application of hurdles overcame the limitations with a single use of EW. Hence, as a progressive and dynamic sanitizing method, hurdle combination of EW with abovementioned treatments is capable of being a novel sanitizer in food processing to overcome the all current drawbacks of other sanitizers.

References

1. CDC (2020) Estimates of foodborne illness in the United States. https://www.cdc.gov/foodborneburden/index.html
2. Rahman SME, Khan I, Oh DH (2016) Electrolysed water as a novel sanitizer in the food industry: current trends and future perspectives. Compr Rev Food Sci Food Saf 15: 471–490
3. Al-Haq MI, Sugiyama J, Isobe S (2005) Applications of electrolyzed water in agriculture & food industries. Food Sci Technol Res 11(2): 135–150
4. Rahman SME, Ding T, Oh DH (2010) Inactivation effect of newly developed low concentration electrolyzed water and other sanitizers against microorganisms on spinach. Food Control 21(10):1383–1387
5. Lopez-Galvez F, Allende A, Gil MI (2021) Recent progress on the management of the industrial washing of fresh produce with a focus on microbiological risks. Curr Opin Food Sci 38:46–51
6. Kim C, Hung YC, Brackett RE (2000) Efficacy of electrolyzed oxidizing (EO) and chemically modified water on different types of foodborne pathogens. Int J Food Microbiol 61(2): 199–207
7. Nikulin V (1977) Use of an electrolyzed sodium chloride solution for disinfection in therapeutic and prophylactic institutions. Sov Med 12:105–108
8. Krivobok N, Gadadymov BV, Nosov V, Ter--Minas'ian G (1982) Quantitative evaluation of the effect of physicochemical and technological factors on the water regeneration process. Kosm Biol Aviakosm Med 16(5):91–93
9. Hricova D, Stephan R, Zweifel C (2008) Electrolyzed water and its application in the food industry. J Food Protect 71(9):1934–1947
10. Zhang J, Yang H (2017) Effects of potential organic compatible sanitisers on organic and conventional fresh-cut lettuce (Lactuca sativa Var. Crispa L). Food Control 72:20–26
11. Zhao L, Zhang Y, Yang H (2017) Efficacy of low concentration neutralised electrolysed water and ultrasound combination for inactivating Escherichia coli ATCC 25922, Pichia pastoris GS115 and Aureobasidium pullulans 2012 on stainless steel coupons. Food Control 73:889–899
12. Ding T, Ge Z, Shi J, Xu YT, Jones CL, Liu DH (2015) Impact of slightly acidic electrolyzed water (SAEW) and ultrasound on microbial loads and quality of fresh fruits. LWT Food Sci Technol 60(2):1195–1199
13. Hao J, Li H, Wan Y, Liu H (2015) Combined effect of acidic electrolyzed water (AcEW) and alkaline electrolyzed water (AlEW) on the microbial reduction of fresh-cut cilantro. Food Control 50:699–704
14. Huang YR, Hung YC, Hsu SY, Huang YW, Hwang DF (2008) Application of electrolyzed water in the food industry. Food Control 19(4):329–345
15. Liu C, Su YC (2006) Efficiency of electrolyzed oxidizing water on reducing Listeria monocytogenes contamination on seafood processing gloves. Int J Food Microbiol 110(2):149–154
16. Monnin A, Lee J, Pascall MA (2012) Efficacy of neutral electrolyzed water for sanitization of cutting boards used in the preparation of foods. J Food Eng 110(4):541–546
17. Lin T, Wang JJ, Li JB, Liao C, Pan YJ, Zhao Y (2013) Use of acidic electrolyzed water ice for preserving the quality of shrimp. J Agric Food Chem 61:8695–8702
18. Al-Holy MA, Rasco BA (2015) The bactericidal activity of acidic electrolyzed oxidizing water against Escherichia coli O157:H7, Salmonella typhimurium, and Listeria monocytogenes on raw fish, chicken, and beef surfaces. Food Control 54:317–321
19. Ding T, Rahman SME, Purev U, Oh DH (2010) Modelling of Escherichia coli O157: H7 growth at various storage temperatures on beef treated with electrolyzed oxidizing water. J Food Eng 97(4):497–403
20. Rahman SME, Park J, Song KB, Al-Harbi NA, Oh DH (2012) Effects of slightly acidic low concentration electrolyzed water on microbiological, physicochemical, and sensory quality of fresh chicken breast meat. J Food Sci 77(1): M35–M41
21. Wang J, Rahman SME, Park MS, Park JH, Oh DH (2012) Modeling the response of Listeria monocytogenes at various storage temperatures in pork with/without electrolyzed water treatment. Food Sci Biotechnol 21(6): 1549–1555
22. Graca A, Abadias M, Salazar M, Nunes C (2011) The use of electrolyzed water as a disinfectant for minimally processed apples. Postharvest Biol Technol 61(2):172–177
23. Ding T, Dong QL, Rahman SME, Oh DH (2011) Response surface modeling of Listeria monocytogenes inactivation on lettuce treated

with electrolyzed oxidizing water. J Food Process Eng 34(5):1729–1745

24. Mansur AR, Oh DH (2015) Combined effect of thermosonication and slightly acidic electrolyzed water to reduce foodborne pathogens and spoilage microorganisms on fresh-cut kale. J Food Sci 80(6):M1277–M1M84
25. Singh S, Shalini R (2016) Effect of hurdle technology in food preservation: a review. Crit Rev Food Sci Nutr 56(4):641–649
26. Rahman SME, Yong-Guo J, Oh DH (2011) Combination treatment of alkaline electrolyzed water and citric acid with mild heat to ensure microbial safety, shelf-life and sensory quality of shredded carrots. Food Microbiol 28(3): 484–491
27. Forghani F, Rahman SME, Park MS, Park JH, Park J, Song KB, Oh DH (2013) Ultrasonication enhanced low concentration electrolyzed water efficacy on bacteria inactivation and shelf life extension on lettuce. Food Sci Biotechnol 22(1):131–136
28. Liu Q, Jin X, Feng X, Yang H, Fu C (2019) Inactivation kinetics of Escherichia coli O157: H7 and Salmonella typhimurium on organic carrot (Daucus carota L.) treated with low concentration electrolyzed water combined with short-time heat treatment. Food Control 106: 106702
29. Rahman SME, Wang J, Oh DH (2013) Synergistic effect of low concentration electrolyzed water and calcium lactate to ensure microbial safety, shelf life and sensory quality of fresh pork. Food Control 30(1):176–183
30. Abadias M, Usall J, Oliveira M, Alegre I, Viñas I (2008) Efficacy of neutral electrolyzed water (NEW) for reducing microbial contamination on minimally processed vegetables. Int J Food Microbiol 123(1–2):151–158
31. Tango CN, Mansur AR, Oh DH (2015) Fumaric acid and slightly acidic electrolyzed water inactivate gram positive and gram negative foodborne pathogens. Microorganisms 3(1): 34–46
32. Park YB, Guo JY, Rahman SME, Ahn J, Oh DH (2009) Synergistic effect of electrolyzed water and citric acid against Bacillus cereus cells and spores on cereal grains. J Food Sci 74(4):185–189
33. Latha C, Sherikar AT, Waskar VS, Dubal ZB, Ahmed SN (2009) Sanitizing effect of salts on experimentally inoculated organisms on pork carcasses. Meat Sci 83(4):796–799
34. Deza M, Araujo M, Garrido M (2007) Efficacy of neutral electrolyzed water to inactivate Escherichia coli, Listeria monocytogenes, Pseudomonas aeruginosa, and Staphylococcus aureus on plastic and wooden kitchen cutting boards. J Food Protect 70(1):102–108
35. Nan S, Yongyu L, Baoming L, Wang C, Cui X, Cao W (2010) Effect of slightly acidic electrolyzed water for inactivating Escherichia coli O157:H7 and Staphylococcus aureus analyzed by transmission electron microscopy. J Food Protect 73(12):2211–2216
36. Forghani F, Park JH, Oh DH (2015) Effect of water hardness on the production and microbicidal efficacy of slightly acidic electrolyzed water. Food Microbiol 48:28–34
37. Mansur AR, Tango CN, Kim GH, Oh DH (2015) Combined effects of slightly acidic electrolyzed water and fumaric acid on the reduction of foodborne pathogens and shelf life extension of fresh pork. Food Control 47: 277–284
38. Picouet PA, Sárraga C, Cofán S, Belletti N, Guàrdia MD (2015) Effects of thermal and high-pressure treatments on the carotene content, microbiological safety and sensory properties of acidified and of non-acidified carrot juice. Food Sci Technol 62:920–926
39. Riganakos KA, Karabagias IK, Gertzou I, Stahl M (2017) Comparison of UV-Cand thermal treatments for the preservation of carrot juice. Innov Food Sci Emerg Technol 42:165–172
40. Iram A, Wang X, Demirci A (2021) Electrolyzed oxidizing water and its applications as sanitation and cleaning agent. Food Eng Rev 13:411–427
41. Iwasawa A, Nakamura Y (1993) Antimicrobial activity of aqua oxidizing water. Clin Bacteriol 20:469–473

Chapter 7

Ultrasound-Assisted Drying of Food

Małgorzata Nowacka and Magdalena Dadan

Abstract

Food is commonly preserved by means of the drying process, which guarantees that the obtained product is durable and can be stored at ambient temperature for a long time. However, drying is a time-consuming process, during which the elevated temperature negatively affects the product quality. To improve the functional and physical properties of dried materials as well as shorten the drying time, ultrasound (US) can be used as a treatment before the drying process or during the process. When ultrasound is generated during pretreatment, the following devices are used: ultrasonic bath, probe, and contact ultrasound. On the other hand, for the ultrasound-assisted drying, special dryers are designed, which are equipped with transducers mounted to the sieve or as airborne ultrasound. Our goal is to provide information on how to conduct the drying process with the assistance of ultrasound using different devices. An overview of the devices used in the available scientific literature is presented.

Key words Ultrasound, Sonication, Drying, Plant tissue, Ultrasound pretreatment, Ultrasound-assisted drying

1 Introduction

In recent years, nonthermal methods are used before or during the drying process. Ultrasound (US) is one of the nonthermal technologies, which may support the diffusion and intensify the mass and heat transfer processes but also improve the quality of the final product, including beneficial alterations of physical properties and increased retention of bioactive compounds of the plant material. The application of ultrasound in the drying process is still limited to the laboratory scale or pilot scale, although the advantages of the use of sonication are very promising [1]. Different studies show that to obtain an acceptable quality of dried material, the sonication process should be conducted with appropriately selected parameters. Thus, the aim of this chapter is to describe the possibility of the use of sonication to support the drying process.

Mohsen Gavahian (ed.), *Emerging Food Processing Technologies*,
Methods and Protocols in Food Science, https://doi.org/10.1007/978-1-0716-2136-3_7,

1.1 Drying

Drying is very often used in food processing and is probably the oldest method of food preservation [1–3]. Drying ensures microbiological safety by reducing the availability of water necessary for the microorganism's development and slowing down or inhibiting chemical reactions and physical changes. At the same time, the mass and volume of the dried materials are reduced during drying, which in turn decreases the cost of packaging, transportation, and storage. Additionally, by selecting the appropriate drying methods and parameters, it is possible to give the products special forms and quality characteristics [4–8]. The purpose of the drying process in the food industry is to obtain a high-quality product, but also, from the economic and environmental point of view, it is important to reduce the energy consumption and overall costs of the process. It is especially important, as based on estimates, around 20% of the energy used in food processing are consumed during drying [9, 10]. One of the difficulties is that the drying process has the greatest variety of equipment among unit operations of process engineering [11, 12]. To minimize energy consumption, and thus reduce expenditures, researchers look for new solutions, e.g., the use of different treatments and preliminary operations [13, 14] as well as different drying methods [15–18]. Ultrasound can be considered as one of these novel techniques used as pretreatment or during the drying process [1].

Convective drying is one of the common methods of food drying. Water from the material is removed by evaporation under the influence of high temperature. This process has low energy efficiency and is characterized by long drying. The operating costs are high due to the increase of energy consumption [17, 19]. The long exposure to elevated temperature causes also irreversible changes in the chemical and physical characteristics of the dried materials. Usually, convective dried products are characterized by low quality due to high degradation of bioactive components, visible color changes, and high shrinkage [20, 21]. Due to this, other technologies and dryer construction are designed by the researchers. Among the different methods of plant tissue drying, the following are predominantly investigated: microwave [18], infrared [22], vacuum [23], freeze-drying [24], and different hybrid methods that combine different types of drying [1, 5, 25].

1.2 Ultrasound

Ultrasound (US) can be defined as an energy generated by sound waves of frequency in the range of 18 kHz–100 MHz, which are not heard by humans [1, 10, 26–28]. Unlike the human species, some animals are naturally endowed with the ability to register ultrasound, such as dogs. Bats have used ultrasound to locate potential prey and determine its size and speed, whereas some sea whales use high-intensity ultrasound to stun their prey prior to capture [29]. With the technological advances of the last century, scientists began to conduct research on ultrasound and its potential usefulness for humans.

Ultrasound, after being emitted by a source, propagates in a sinusoidal wave and transmits mechanical energy to the structure of a medium other than air, thus causing it to vibrate. The vibrations take the form of compression and expansion, thanks to which the intermolecular spaces in the medium are, respectively, smaller or larger [30]. Ultrasound can propagate in solid, liquid, and gaseous media, and the condition for wave propagation is non-zero mass. Due to the high absorption of acoustic wave energy by air, the spectrum of application of air ultrasound as a medium is limited and difficult to develop [31].

The use of ultrasound is related to its frequency and intensity. Two types of waves can be distinguished depending on their intensity [10, 18, 26–29, 32–34]:

- **Low-intensity ultrasound** with frequencies higher than 100 kHz and intensity below 1 W/cm^2 is used in cosmetology, metrology, diagnostic medicine (ultrasonography), and determination of food composition;
- **High-intensity ultrasound** (high-power ultrasound) with a frequency between 18 and 100 kHz (usually in the range of 20–40 kHz) and intensity above 1 W/cm^2 is used to modify the physical properties of plant tissues and to intensify chemical reactions and mass transfer-based processes.

The low-intensity ultrasound is generally safe and does not cause any chemical or physical changes to the material, which it passes through. Thus, it is used for noninvasive analytical techniques in ultrasonography or to determine the chemical composition, particle size (such as air in foam or fat in milk), product quality (e.g., fruit ripeness), and physical properties of food. Furthermore, devices such as thickness and flow meters, which base their operating principles on ultrasound, are widely used in the industry for monitoring of process parameters as they can be easily adapted to online monitoring of food processing and can be applied non-intrusively for concentrated and optically opaque systems [29, 33, 35].

The high-intensity ultrasound, also called power ultrasound, causes physical and chemical changes in the food. Power ultrasound is used to support different processes in food technology such as drying, freezing, osmotic dehydration, extraction, cooking, filtration, homogenization, emulsification, crystallization, disinfection, enzyme inactivation, degassing, etc. [36–39]. Due to the abovementioned applications, the chapter focuses on the high-intensity acoustic waves.

1.3 Ultrasound Mechanisms

High-power and high-intensity ultrasounds have the ability to induce changes in the propagation medium, and ultrasonic waves are absorbed by the plant materials, which results in thermal, mechanical, and physicochemical effects. The use of ultrasound in technology causes a number of phenomena such as cavitation and effects accompanying cavitation (microstreaming, microjetting), "sponge effect," and absorption of acoustic energy [1, 10, 28, 36].

The cavitation effect is connected with the propagation of the acoustic waves in liquids as well as solids, which contain moisture. The phenomenon of cavitation is the formation, growth, and implosion of gas microbubbles, which is caused by local drops in liquid pressure during sonication. This causes damage to cells and facilitates the removal of water and also causes turbulence, microjets, shock waves (generation of very high pressures up to 50–100 MPa and temperature up to 5500 °C), and shear stresses in the fluid medium. The destruction of cell walls and membranes resulted in inactivation of microorganisms, mainly bacteria and yeasts. Microjets intensify the heat and mass transfer. In addition, the breakdown of water molecules causes the formation of hydroxyl radicals, superoxide radicals, and free hydrogen atoms. During radical interaction with the material, it comes to the initiation of many chemical reactions. The occurrence of the phenomenon of cavitation is associated with many factors which are related, among others, with the product properties (viscosity, surface tension), the properties of the ultrasound (intensity of ultrasonic waves, frequency), and the environmental conditions (temperature, pressure) [1, 28, 36, 37, 40, 41].

"Sponge effect" is based on the compressions and depressions related to the acoustics wave passage through the solid material. This causes mechanical stress and results in microcracks, and microchannels are created in the structure [42–45].

Apart from the physical consequences mentioned above, cavitation also has chemical (sonochemical) effects. The bubbles that form during the cavitation phenomenon interact with water molecules in plant tissue. This generates the appearance of free radicals that can intensify chemical reactions and can also be the starting point for microstreaming. Hydroxyl OH• radicals and hydrogen atoms are formed in aqueous solutions as a result of the action of high temperature and pressure accompanying the collapse of the bubbles [45]. The amount of free radicals formed depends on the bubble size and is higher at lower ultrasound frequencies. Generation of free radicals is often an undesirable effect due to their possible oxidizing effects on phenolic compounds and other food ingredients. The solution is to add free radical scavengers such as ascorbic acid or ethanol to the medium [10, 33].

The formation of free radicals may also affect the quality of foodstuffs by reducing the antioxidant capacity [46, 47]. An important effect of the propagation of ultrasound at appropriate frequencies and intensities is the inactivation of the biochemical activity of enzymes in plant tissues. The formation of stresses and high temperatures during cavitation can lead to the breaking of hydrogen bonds and van der Waals interactions in the polypeptide chains, which results in the modification of the secondary and tertiary structures of the protein. This usually means a loss of enzyme activity. It is also possible to activate enzymes by sonication. The physical effect of cavitation in the form of microstreaming and point turbulence translates into the biochemical availability of the enzyme to the substrate and vice versa. Sometimes the enzyme becomes immobilized, which may also have a positive effect on its availability to substrates. A purely biochemical effect is understood as the induction of changes in living tissue cells by ultrasound, leading to an increase in the production of certain enzymes [47].

As a rule, each fruit reacts differently to sonication [48]. The effects of ultrasound propagation through plant tissue are influenced by its structural attributes, in particular porosity, which is closely related to the expansion of the structure. The parameters of the emitted sound waves are also important, that is, acoustic energy and frequency. Temperature is considered the most important controllable parameter of the medium [31].

2 Materials

2.1 Sample Preparation

For the drying process, the plant material is usually prepared in pieces. Before the pretreatment, the raw material has to be cleaned with tap water [20], sometimes with the addition of different chlorine disinfectants, e.g., sodium and calcium hypochlorite [49], and cut into chosen shapes (cubes, slices, tissue cut with specific dimensions). Usually, the shape proposed by researchers is related to the type of plant tissue or to the simple and quick preparation material in the same shape. Fruits or vegetable flesh are damaged during cutting, which led to enzymatic reactions and degradation of bioactive compounds [50, 51]. In some cases, there is necessary to peel the skin before cutting [52]. The obtained samples are placed in a device generating ultrasound (pretreatment) or in a dryer in which the ultrasound works during drying (ultrasound-assisted drying).

2.2 Ultrasound Equipment

For pretreatment and ultrasound-assisted drying, different ultrasonic systems are used (Table 1). However, every type of ultrasound system is built from electrical power generator, transducers, and emitters (*see* **Note 1**).

To conduct the pretreatment before drying as well as sonication during drying, usually the following systems are utilized:

Table 1
Characteristics of different ultrasonic systems

Type of ultrasonic system	The characteristic of the system				
	Type	Frequency [kHz]	Effects	Other feature	Application
1 Ultrasonic bath	Indirect sonication.	20–40	The transducers produce high-intensity waves, which result in compression and rarefaction as well as production of cavitation's bubbles	–	Pretreatment
2 Ultrasonic probe	Direct sonication.	20–24	The ultrasound causes cavitation in the medium, and waves are transferred to the solid materials	The horns/probes have different diameters and acoustic power densities	Pretreatment, ultrasound-assisted drying
3 Contact ultrasound	Direct sonication.	20–28	Mainly the "sponge effect" and in solid materials containing high amount of moisture, the cavitation may also occur inside the tissue	Transducers can be mounted to sieves/trays/plates on which the material is placed (pretreatment, US-assisted drying) or the material is in close contact with the transducer (US-assisted drying)	Pretreatment, ultrasound-assisted drying
4 Airborne ultrasound technology	Indirect sonication.	20–22	Mainly the "sponge effect," acoustic pressure variation, as well as microjetting and microstreaming occur causing microchannels formation	The ultrasound is emitted as radiation through the air by plate transducers of different shape	Pretreatment, ultrasound-assisted drying

- Ultrasonic bath—the ultrasound is generated by transducers (one or more in the tank), which are placed at the walls or base of the ultrasonic tank. Transducer changes the electrical energy into sound energy by mechanical vibrations at ultrasonic frequencies. Ultrasonic waves from the transducers are propagated through the liquid (e.g., water), which must fill the device. It is important that the liquid should be filled up to the line or to at least half of its height to maintain the intensity of waves and frequency (if it is not marked by a producer). Ultrasound baths usually work at a frequency in a range from 20 to 40 kHz and produce high-intensity waves (*see* **Note 2**). Due to this, cavitation bubbles are created, which grow and rapidly collapse generating mechanical and chemical energy [53–55].
- Ultrasonic probe—the sonotrode is attached to the ultrasonic processor through the probe/horn. It is required to choose the diameter of the probes and set the amplitude, cycle, and time. Sonotrode should be placed at an optimum working height in a beaker, and the submerged depth can be different. The sonotrode generates cavitation in the liquid, and ultrasonic waves are transferred to the plant tissue dipped in the liquid. During the sonication, the temperature of the sample is increasing rapidly (*see* **Note 3**); thus the ultrasonic probe should be used shortly with small volume samples [53, 55, 56].
- Contact ultrasound—in the case of pretreatment, the sonotrode can be attached to the sieve/tray with material as, e.g., ring sonotrode, or can be mounted to the sieve/tray/plate. In the case of US-assisted drying, the transducers can also be coupled near the material [45, 57, 58]. The dryer can be equipped with the cooling system. Mainly the power intensity and the vibrational amplitude are controlled [45, 53]. Usually, frequencies around 20–28 kHz are used for this purpose [45, 57]. The changes in the material are due to the sponge effect and increase of the temperature of a material [59, 60].
- Airborne ultrasound—the transducer is not in contact with the material. The plate piezoelectric transducer emits ultrasound as radiation to the material. Different types of transducers and their shapes are utilized (e.g., flat plate, stepped circular plate, stepped rectangular plate, grooved plate, cylindrical plate, etc.). The ultrasound frequency is depended on the transducer. For instance, the circular stepped-plate radiator works at 10–40 kHz, whereas the rectangular stepped-plate radiator works at 7–20 kHz. Usually, for drying purposes, the frequency around 20–22 kHz is used. Ultrasound causes acoustics pressure variation, compression and rarefaction, microstreaming, and microjetting, which enhance the drying rate [13].

2.3 Other Materials

Due to the fact that ultrasound spreads well in liquids, various types of media can be used. The most commonly used medium is tap water. Distilled water may also be used, but due to the expected industrial applications of sonication (e.g., during washing before drying), tap water is preferable [26, 48]. Additionally, various solutions may be used such as juices, solutions containing vitamins, osmotic solutions to obtain higher quality of the final product (*see* **Note 4**), or ethanol to enhance the drying process [20, 50, 61–63].

3 Methods

3.1 Sonication in an Ultrasonic Bath

The most popular method for ultrasound pretreatment is the use of an indirect mode by means of an ultrasonic bath [32]. It is crucial to check the power and frequency of the ultrasonic bath, which can be found in the manufacturer's instructions. Some ultrasonic baths operate in a specific frequency range that does not guarantee constant parameters.

The processing usually is carried out in a glass beaker (preferably made of thin glass) in which previously prepared raw material followed by liquid (e.g., tap or distilled water) is added (Fig. 1) (*see* **Note 5**). The mass and temperature of the samples should be measured before and after the treatment. Moreover, it should be taken into account that if the double portion of samples is prepared in two beakers (the same type and volume) and put into the ultrasonic bath, the experiments should be conducted always in the same way (with two beakers) (*see* **Note 6**). This is important to assure the same ultrasound intensity which goes through the material in each experiment (*see* **Note 7**). The ratio of raw material to water should be at least 1:4 [64, 65] or higher, e.g., 1:5 (*see* **Note**

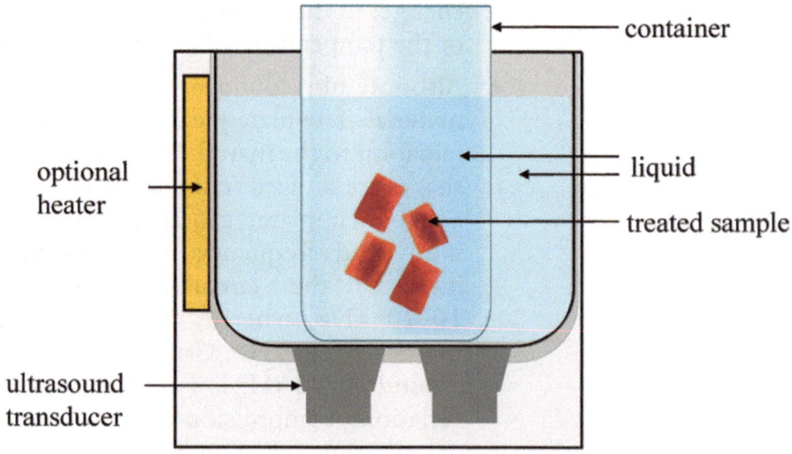

Fig. 1 Schematic diagram of the sonication process with ultrasonic bath (based on [55])

8), to cover the samples with the liquid. If the sample flows to the surface, it is necessary to place a mesh to assure a full immersion of the material in the medium, due to the fact that ultrasounds are transferred through the liquid and reach the sample dipped in the liquid [52].

After placing the beaker in the ultrasound bath, the distilled water should be filled up to the line marked in the bath (*see* **Note 9**) or at least half of the bath volume (if it is not marked by the producer). The water volume should be in a similar line as the liquid in the beaker. It should be noted that the size of the beaker and the amount of water in the bath should be selected in that way to avoid turning over the beaker during processing.

The treatment can be also performed without the use of a beaker, but it should be remembered to select the appropriate ratio of the raw material to the liquid medium. Moreover, in some cases, the sonication can be conducted with vacuum-packed samples. Due to this, mostly the sponge effect will occur during sonication, and the effect of cavitation will be reduced [18, 66].

During treatment, the temperature of both medium and material increases with increasing time of sonication, so this parameter should be controlled. Usually, the time is set in the range of 5–30 min, but also shorter (e.g., 3 min or less) and longer (e.g., 45, 60 min) treatment might be conducted. It is recommended that the sonication process should be relatively short—as short as possible but sufficient to obtain the required effect (e.g., reduction of drying time, higher extractivity of bioactive components, etc.), as it is more economical. Furthermore, longer sonication time contributes to the formation of more OH^- and H^+ radicals [10, 28]. Moreover, the higher temperature of the sonication treatment might be used for some purposes with the heater, which is usually available in the ultrasonic bath.

The samples after the sonication are drained with a sieve, or the excess water is removed by a filter paper, and the mass and temperature are measured after the process.

Different equations are used to calculate ultrasonic power (acoustic intensity), among which the following are most commonly mentioned.

1. The power intensity (acoustic intensity) can be determined calorimetrically by recording the temperature increase with the time of ultrasound application with the following equation [67]:

$$P = MC_p \frac{dT}{dt} \tag{1}$$

   where P is the ultrasonic power (W); M is the mass of the solution (kg); dT/dt is an increase of temperature (K/s); and C_p is the heat capacity of the solution (J/kg K).

2. Acoustic intensity can be determined by dividing the ultrasonic power by the emitting surface of the probe in W/m^2 or W/cm^2 [68]:

$$P = \frac{P_s}{A} \tag{2}$$

where P is the ultrasonic power (W/m^2); P_s is the power of sonotrodes distributed over surface area (W); and A is the area of sonotrodes through which the waves are transmitted (m^2).

3. Acoustic intensity can be determined by dividing the ultrasonic power by the mass of the treated sample in W/g [69]:

$$P = \frac{P_s}{m} \tag{3}$$

where P is the ultrasonic power (W/g); P_s is the power of sonotrodes distributed over surface area (W); and m is the mass of the treated sample (g).

Furthermore, the specific energy consumption E (kJ/kg) might be calculated as follows [70]:

$$E = \frac{P\tilde{n}t}{m} \tag{4}$$

where E is the specific energy consumption (kJ/kg); t is the total time (s); m is the sample weight (kg); and P is the power of a generator (W).

3.2 Sonication with the Probe

The use of sonication on the probes/horn is also popular for food pretreatment before drying [32]. The probe system operates as a direct sonication into the liquid, and it is dedicated to small volumes. It causes high-power acoustic cavitation. The frequency of that system is around 20–24 kHz [32, 55, 67]. In comparison to the ultrasound bath, this system provides higher power intensity, approximately 100 times higher [67, 71], as the ultrasonic waves are transmitted to the sample by high-power probe of a small circle area normal to the direction of ultrasound propagation (*see* Eq. 2) [55].

The ultrasonic probe consists of an ultrasonic generator connected to an ultrasonic processor (transducer) with the booster and equipped with the probe or horn. The ultrasonic processor is electrically connected with the ultrasound generator (Fig. 2). The ultrasonic processor generates longitudinal mechanical vibrations with a specific frequency. What is important, the sonotrode must be firmly connected with the horn of the ultrasonic processor, in order to work properly. Vibrations are amplified by the sonotrode fitted to the horn and transferred via its end face to the medium. Due to the high acoustic power and the small volume of sample, the rapid increase of the medium and material's temperature must be taken into account [55]. The horn and sonotrode can heat even up to

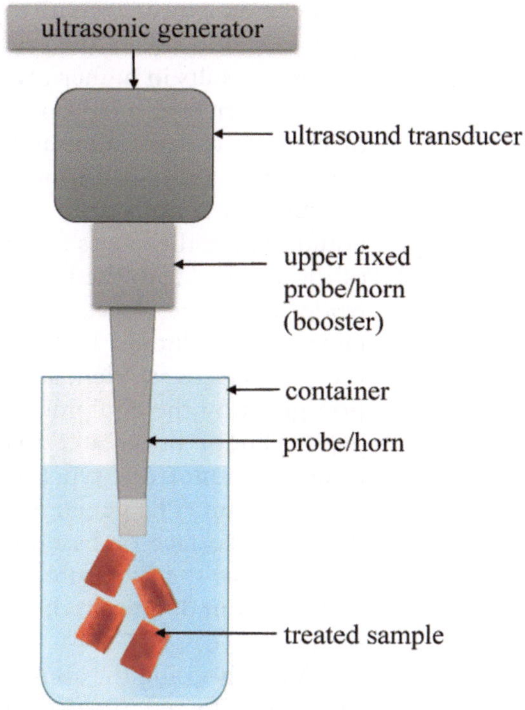

Fig. 2 Schematic diagram of the sonication process with an ultrasonic probe (based on [54])

100 °C; thus cooling systems are sometimes required to avoid overheating of the material (*see* **Note 10**).

Before starting the sonication, the proper parameters have to be chosen for amplitude, cycle, and time of sonication. The ultrasonic probe operates with oscillation amplitude between 20 and 100%. Furthermore, the pulse control mode (duty cycle) can be used in the range from 0.1 to 1, which means that the ultrasound is switched on and switched off in a specific time, e.g., duty cycle set on 0.6 means that the ultrasound is propagated for 0.6 s and the break (without ultrasound generation) lasts 0.4 s. The time of treatment is shorter in comparison to the treatment in an ultrasound bath, and it is limited by the increase of the temperature of the sample [32].

The acoustic power density depends on the tip of a probe, where its shape determines the amount of amplification. Thus, the ultrasound intensity is set by choosing the shape of the sonotrode. In the manual, the manufacturers of the ultrasonic probe present the characteristics of the tips. In general, the acoustic power density and amplitude decrease with the end face area increase. Tip diameter from 3 to 40 mm can generate acoustic power density in the range of 12–460 W/cm^2. Generally, acoustic power density used before drying is between 100 and 600 W, but different devices can generate power even up to 900 W [72]. Moreover, the shape of the sonotrode affects the cavitation intensity, and the smaller end face area results in higher cavitation.

Furthermore, each sonotrode is designed for a specific volume of material and medium and requires an appropriate penetration depth. The abovementioned should be taken into account when selecting the sonotrode type. The sonotrode has to be dipped in the liquid, where the waves are transferred; however, different depths of the sonotrode in the liquid give diverse results. Thus, the sonotrode should be placed at an optimum working height in a container, which should be chosen on the basis of the literature or pretest. Moreover, scientists have to take into account the information given by the equipment producer in the instruction that the depth cannot be greater than the maximum submerged of the respective sonotrode type; usually, this will be in the range from 20 to 90 mm. The penetration depth is given as the distance from the water surface or base of the beaker to the tip of sonotrode [32, 67]. In the literature, the range of the sonotrode dip in the liquid is from 30 mm to the height equal to 5 mm from the base of the beaker [67, 72].

After choosing the abovementioned parameters, the sonication can be conducted. The samples are placed in the liquid, and the ultrasound causes cavitation in the medium, and waves are transferred to the plant material.

3.3 Contact Ultrasound

The contact ultrasounds work in a direct way, through the net or plate; the ultrasound waves are distributed to the material. The ultrasound generator is connected to the ultrasonic processor, and the waves are transferred via, e.g., ring sonotrode (Fig. 3), which causes vibrations of the sieve or plate. The contact ultrasound operates at frequencies in the range of 20–28 kHz, and the power

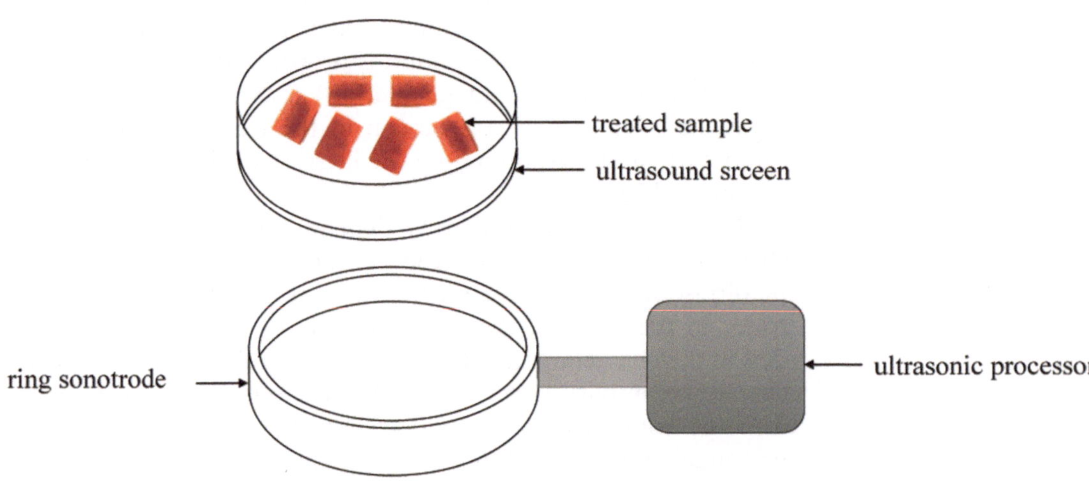

Fig. 3 Schematic diagram of the sonication process with contact ultrasound (based on [57])

is up to 300 W. The intensity of the contact ultrasound processing is defined by the regulation of the amplitude, which is in the range of 20–100% [45, 57, 59].

The samples are placed on the sieve/tray, and the ultrasound generator of the specified amplitude is turned on. Due to the fact that ultrasounds are transferred by the net/tray, the material is exposed to the air (not immersed in the medium as in the case of bath or probe) [60]. In general, in the material treated with contact ultrasound, the "sponge effect" is the predominant phenomenon. However, in solid materials which contain moisture (e.g., fresh fruits and vegetables), the cavitation may also occur inside the tissue [28, 60]. Nevertheless, this effect is rather limited.

Due to the availability of oxygen during contact ultrasound processing, there would be a negative influence on color and bioactive compounds content [45, 60]. Contact ultrasound is more often used for ultrasonic-assisted drying.

3.4 Drying of the Sonicated Samples

After the sonication is conducted with the use of the ultrasonic bath, ultrasonic probe, or contact ultrasound with specifically chosen parameters, the samples are weighed and placed in a dryer. Drying process is monitored by recording of the mass and temperature changes of both air and a material. Depending on the type of the dryer used, other parameters have to be properly chosen, e.g., airflow for convective or infrared drying, vacuum, and temperature of the shelf for freeze-drying, microwave power for microwave drying, the level of vacuum for vacuum drying, the humidity of air, etc. (*see* **Note 11**). As the process control is done in the same way as in the case of US-assisted drying, the main parameters used for this purpose are explained in Subheading 3.5. Furthermore, it is required to optimize the process considering the type of tissue and different parameters of sonication (*see* **Note 12**).

3.5 Sonication Applied During Drying

Ultrasound-assisted drying is generally more effective in enhancing the process than ultrasound applied before drying [10]. However, from an investment point of view, the sonication applied during drying requires the modification of an existing technological lines or the construction of a dedicated type of dryer. The key elements of such a dryer are the generator and transducer, whereas the rest part of dryer usually is typical for a specific type of dryer. For instance, the transducer can be mounted to the belt dryer, freeze-dryer, chamber or vacuum dryer, microwave dryer, heat pump dryer, etc. [57, 58, 73–75]. Based on the transducer type and its way of mounting, the ultrasound-assisted drying can be carried out by means of:

- Contact ultrasound (Fig. 4)—in which the transducer is mounted to the sieve/tray on which the material is placed or the material is in close contact with the transducer. The material

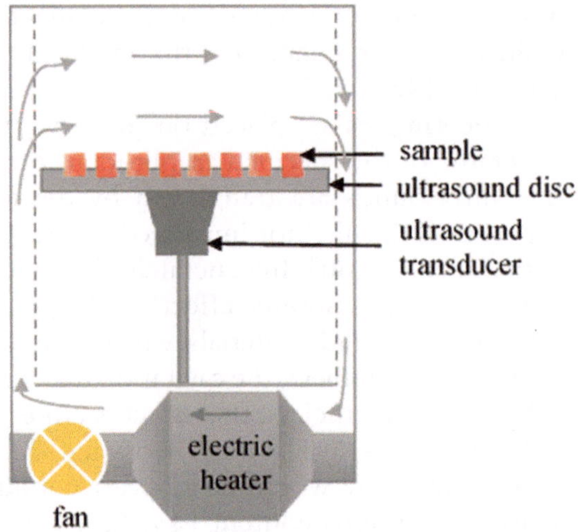

Fig. 4 Schematic diagram of contact ultrasound-assisted dryer (based on [45])

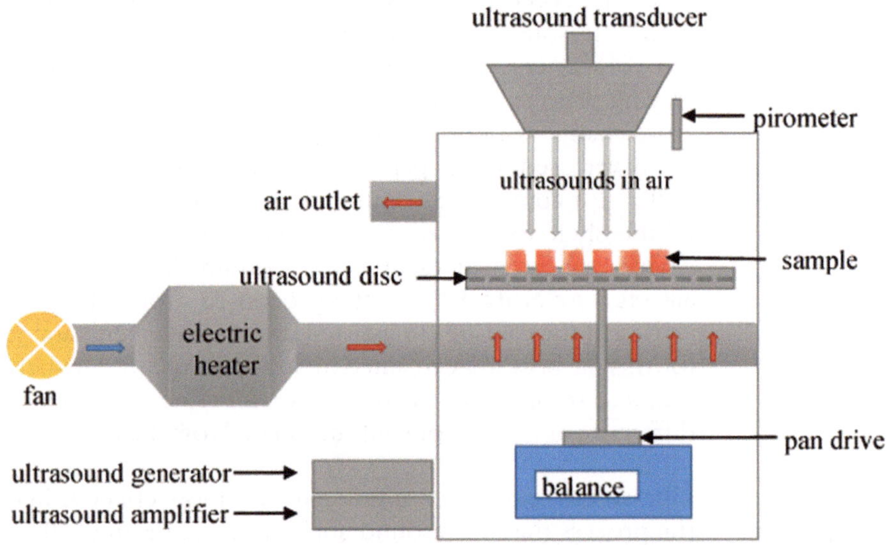

Fig. 5 Schematic diagram of airborne ultrasound-assisted dryer (based on [74])

may be heated to a great extent due to the absorption of the acoustic waves, the expansion and contractions generated by the piezoelectric transducer, as well as the cavitation of liquid in the material causing the bond water to evaporate more easily. Therefore, the material temperature should be controlled [13, 45, 58].

- Airborne ultrasound (Fig. 5)—in which ultrasounds are transmitted through the air. According to the Charoux et al. [13], the airborne ultrasound system dedicated for food drying consists of a stepped-plate transducer and a flat plate, which has a function of both a sample holder and a reflector.

The utilization of the noncontact method of sonication (airborne ultrasound) in drying enhancement is rather limited. There are two main reasons for that situation. Firstly, the acoustic energy is attenuated by a gas medium. Secondly, the large mismatches of the acoustic impedance of the air and the tune transducers and between air and the dried material cause the poor energy propagation of airborne ultrasound [13, 56]. For instance, the acoustic impedance of air is 400 kg/(m^2 s) (at standard temperature and pressure), while for water it is $1.48 \cdot 10^6$ kg/(m^2 s) (at 20 °C) [13]. This is not the case with a contact sonication system for which the acoustic impedance is matched and the ultrasound can deeply penetrate the material. As a consequence, the contact ultrasound system is more effective in enhancement of the drying process [56].

Generally, in order to carry out the ultrasound-assisted drying, the material placed on sieves or trays is loaded into the prepared dryer (e.g., heated to the assumed air temperature) (*see* **Note 13**). During the process, usually the mass and the temperature of the material are recorded. The ultrasound can be regulated via ultrasound power, duty cycle (intermittent mode), or amplitude [45, 57, 59, 60]. In the case of contact systems, the control of the temperature of the material is especially important for the prevention of thermolabile nutrient degradation. Based on the literature data, the power of ultrasound is depended on the type of material and other parameters' set, like air temperature, far-infrared radiation, microwave power, etc. [45, 73]. Usually, when other source of power is used, the ultrasound power should be limited to protect against burning and assure a good quality of a material [45]. In general, the ultrasound power is in the range of 25–110 W in the case of contact sonication [45, 57]. Similar levels of power are also used for airborne ultrasound [13].

The drying process usually is performed until the constant mass is obtained. The end point of the process is established based on the MR factor (moisture ratio) either on the assumed time (which was previously checked that guarantees the constant mass of a material).

When the humidity of the air is not controlled during the drying, the moisture ratio (MR) is calculated as follows [76]:

$$\mathrm{MR} = \frac{M_\tau}{M_0} \tag{5}$$

where M_τ is the water content in time of the process (kg water/kg dry matter) and M_0 is the initial water content (kg water/kg dry matter).

From the scientific point of view, also the intensity of the drying process can be controlled based on the MR (drying kinetics presented as MR in time of the process) or the effective water diffusion coefficient (D_{eff}). The D_{eff} is computed based on the simplified Fick's second law, which can be different, dependent on the shape of the material. For instance, when assuming infinite plate, the equation is as follows [18]:

$$MR = 8/\pi^2 \tilde{n} \exp\left(-\pi^2 \tilde{n} D_{\mathrm{eff}} \tilde{n} \tau / 4 \tilde{n} L^2\right) \quad (6)$$

where τ is the time (s); D_{eff} is the coefficient of diffusion (m^2/s); and L is the half of thickness (m).

Water diffusivity can be estimated using different software (like Table Curve).

The more intense process is when the drying time is shorter (higher drop of the MR in the drying kinetics chart) and when the D_{eff} is higher.

4 Notes

1. When the ultrasound is on, wear protective headphones, because unpleasant and loud noise spreads. To conduct experiments with ultrasound, it is good to have a soundproofing room or box.
2. It is better to use an ultrasound bath with a lower frequency (around 20 kHz) due to the lower temperature rise during processing. If the temperature during the experiment rises above 3–4 °C, the cooling system should be turned on. Please take into account that usually commercial-made ultrasound baths do not have a cooling system. Thus, it should be chosen the ultrasound bath with the cooling system.
3. If you work with a heat-sensitive material, try to avoid prolonged treatment (in particular by probe or contact systems) and drying at high ultrasound power. Consider the use of ultrasonic bath pretreatment or airborne ultrasound drying. Control the temperature of the material during processing.
4. As a medium in ultrasound treatment (liquid in a beaker), instead of water which washes out water-soluble ingredients, it is preferable to use different solutions such as juices, solutions containing vitamins, and osmotic solutions to obtain a higher quality of the final product.
5. After putting the sample in the medium, start ultrasound treatment as soon as possible.
6. If you don't have an ultrasonic bath with cooling system, pay attention to replace the water in the bath before the next experiment. The water should be of the same initial temperature; distilled water is preferable in the bath.
7. When sonicating another sample (repetitions or longer treatment with the same or other parameters, e.g., frequency and power), the same conditions should be ensured, e.g., the same weight of the sample, the same volume of water, and the same

initial temperature of water, to ensure similar ultrasonic treatment conditions. Moreover, in an ultrasound bath, the water should be filled in with the same amount.

8. Check if the ratio of the plant material to water is 1:4 and water covers the samples.
9. For ultrasound treatment in the bath, check if the water line in the beaker with the samples is in the same line with water in the ultrasound bath, and check if the beaker will not fall over. The best is to run the first experiment as a trial to check if everything works properly.
10. The temperature of the sample and medium treated with an ultrasound probe increases fast, so cooling is required. The beaker with icy water can be used as a cooling system.
11. When you dry the material for the first time with selected processing parameters, it is better to use lower ultrasound power, lower air temperature, etc., especially if you also use other sources of power (like microwave power, infrared radiation, or high temperature). This will prevent the product from burning. Next, if the drying intensity is too slow, try to use a higher power or more intense parameters.
12. The optimization of the process, including the sonication treatment and drying process, is recommended for specific food materials.
13. Be sure to place the material for drying into a dryer when the assumed conditions are achieved (e.g., air temperature).

5 Conclusions

Generally, the ultrasound treatment used before drying as well as ultrasound-assisted drying leads to obtain the dried material in a shorter time with a better preserved color and better quality of the bioactive compounds. However, it is required to optimize the process considering the type of tissue and different parameters of sonication. From the industrial point of view, the evaluation of the efficiency of the process is essential. The current summarizing is helpful in the assessment of possible sonication solutions and protocols.

References

1. Dadan M, Nowacka M, Wiktor A et al (2021) Ultrasound to improve drying processes and prevent thermolabile nutrients degradation. In: Barba FJ, Cravotto G, Chemat F et al (eds) Design and optimization of innovative food processing techniques assisted by ultrasound. Elsevier Academic Press, London, pp 55–110
2. Tylewicz U, Nowacka M, Rybak K et al (2020) Design of healthy snack based on kiwifruit. Molecules 25:1–13

3. Rybak K, Samborska K, Jedlinska A et al (2020) The impact of pulsed electric field pretreatment of bell pepper on the selected properties of spray dried juice. Innov Food Sci Emerg Technol 65:102446
4. Ozcelik M, Ambros S, Morais SIF et al (2019) Storage stability of dried raspberry foam as a snack product: effect of foam structure and microwave-assisted freeze drying on the stability of plant bioactives and ascorbic acid. J Food Eng 270:109779
5. Kowalska H, Marzec A, Kowalska J et al (2020) The use of a hybrid drying method with pre-osmotic treatment in strawberry bio-snack technology. Int J Food Eng 16:1–14
6. Bai-Ngew S, Therdthai N, Dhamvithee P (2011) Characterization of microwave vacuum-dried durian chips. J Food Eng 104: 114–122
7. Singh A, Sarkar J, Rekha Sahoo R (2020) Experimentation on solar-assisted heat pump dryer: thermodynamic, economic and exergoeconomic assessments. Sol Energy 208: 150–159
8. Jeszka-Skowron M, Zgoła-Grześkowiak A, Stanisz E et al (2017) Potential health benefits and quality of dried fruits: goji fruits, cranberries and raisins. Food Chem 221:228–236
9. Stojceska V, Atuonwu J, Tassou SA (2019) Ohmic and conventional drying of citrus products: energy efficiency, greenhouse gas emissions and nutritional properties. Energy Procedia 161:165–173
10. Witrowa-Rajchert D, Wiktor A, Sledz M et al (2014) Selected emerging technologies to enhance the drying process: a review. Dry Technol 32:1386–1396
11. Chua KJ, Chou SK (2005) New hybrid drying technologies. In: Sun D-W (ed) Emerging technologies for food processing. Elsevier Academic Press, London, pp 535–551
12. Chua KJ, Chou SK (2003) Low-cost drying methods for developing countries. Trends Food Sci Technol 14:519–528
13. Charoux CMG, Ojha KS, O'Donnell CP et al (2017) Applications of airborne ultrasonic technology in the food industry. J Food Eng 208:28–36
14. Wiktor A, Iwaniuk M, Śledź M et al (2013) Drying kinetics of apple tissue treated by pulsed electric field. Dry Technol 31:112–119
15. Celejewska K, Mieszczakowska-Frąc M, Konopacka D et al (2018) The influence of ultrasound and cultivar selection on the biocompounds and physicochemical characteristics of dried blueberry (Vaccinium corymbosum L.) snacks. J Food Sci 83:2305–2316
16. Wang H, Zhang M, Mujumdar AS (2014) Comparison of three new drying methods for drying characteristics and quality of shiitake mushroom (Lentinus edodes). Dry Technol 32:1791–1802
17. Kaveh M, Abbaspour-gilandeh Y, Nowacka M (2021) Chemical engineering and processing—process intensification comparison of different drying techniques and their carbon emissions in green peas. Chem Eng Process Process Intensif 160:108274
18. Sledz M, Wiktor A, Nowacka M et al (2017) Drying kinetics, microstructure and antioxidant properties of basil treated by ultrasound. J Food Process Eng 40:1–13
19. Torki-Harchegani M, Ghanbarian D, Ghasemi Pirbalouti A et al (2016) Dehydration behaviour, mathematical modelling, energy efficiency and essential oil yield of peppermint leaves undergoing microwave and hot air treatments. Renew Sust Energ Rev 58:407–418
20. Dadan M, Nowacka M (2021) The assessment of the possibility of using ethanol and ultrasound to design the properties of dried carrot tissue. Appl Sci 11:689
21. Nowacka M, Śledź M, Wiktor A et al (2012) Physical and chemical properties of microwave dried food products. Food Sci Technol Qual 19:5–20
22. Witrowa-Rajchert D, Rzaca M (2009) Effect of drying method on the microstructure and physical properties of dried apples. Dry Technol 27:903–909
23. Bialik M, Wiktor A, Witrowa-Rajchert D et al (2020) Osmotic dehydration and freezing as a suitable pretreatment in the process of vacuum drying kiwiberry: drying kinetics and microstructural changes. Int Agrophys 34:265–272
24. Kowalska J, Kowalska H, Marzec A et al (2018) Dried strawberries as a high nutritional value fruit snack. Food Sci Biotechnol 27:799–807
25. Aktaş M, Khanlari A, Amini A et al (2017) Performance analysis of heat pump and infrared–heat pump drying of grated carrot using energy-exergy methodology. Energy Convers Manag 132:327–338
26. Nowacka M, Wiktor A, Śledź M et al (2012) Drying of ultrasound pretreated apple and its selected physical properties. J Food Eng 113: 427–433
27. Dadan M, Rybak K, Wiktor A et al (2018) Selected chemical composition changes in microwave-convective dried parsley leaves affected by ultrasound and steaming pre-treatments—an optimization approach. Food Chem 239:242–251

28. Kentish S, Ashokkumar M (2011) The physical and chemical effect of ultrasound. In: Feng H, Barosa-Canovas GV, Weiss J (eds) Ultrasound technologies for food and bioprocessing. Springer, New York, pp 1–12
29. McClements DJ (1995) Advances in the application of ultrasound. Trends Food Sci Technol 6:293–299
30. Hecht E (1996) Physics: calculus. Brooks, Pacific Grove
31. García-Pérez JV, Cárcel JA, Riera E et al (2009) Influence of the applied acoustic energy on the drying of carrots and lemon peel. Dry Technol 27:281–287
32. Bromberger Soquetta M, Schmaltz S, Wesz Righes F et al (2018) Effects of pretreatment ultrasound bath and ultrasonic probe, in osmotic dehydration, in the kinetics of oven drying and the physicochemical properties of beet snacks. J Food Process Preserv 42:1–9
33. Awad TS, Moharram HA, Shaltout OE et al (2012) Applications of ultrasound in analysis, processing and quality control of food: a review. Food Res Int 48:410–427
34. Nowacka M, Fijalkowska A, Wiktor A et al (2017) Changes of mechanical and thermal properties of cranberries subjected to ultrasound treatment. Int J Food Eng 13:1–9
35. Al KF, Zhou J, Wang M et al (2021) Mind the gap in the knowledge of the potential food applications of ultrasound based on its mechanism of action. In: Barba FJ, Cravotto G, Chemat F, Lorenzo Rodriguez JM, Munekata PES (eds) Design and optimization of innovative food processing techniques assisted by ultrasound. Elsevier Academic Press, London, pp 1–13
36. Nowacka M, Dadan M, Tylewicz U (2021) Current applications of ultrasound in fruit and vegetables osmotic dehydration processes. Appl Sci 11:1–22
37. Ashokkumar M (2015) Applications of ultrasound in food and bioprocessing. Ultrason Sonochem 25:17–23
38. Thane C, Reddy S (1997) Processing of fruit and vegetables: effect on carotenoids. Nutr Food Sci 97:58–65
39. Paniwnyk L (2017) Applications of ultrasound in processing of liquid foods: a review. Ultrason Sonochem 38:794–806
40. Dolatowski ZJ, Stadnik J, Stasiak D (2007) Applications of ultrasound in food technology. ACTA Sci Pol 63:89–99
41. Kapturowska A, Stolarzewicz I, Chmielewska I et al (2011) Ultrasounds—a tool to inactivate yeast and to extract intracellular protein. Food Sci Technol Qual 4:160–171
42. Miano AC, Rojas ML, Augusto PED (2019) Structural changes caused by ultrasound pretreatment: direct and indirect demonstration in potato cylinders. Ultrason Sonochem 52: 176–183
43. Rajewska K, Mierzwa D (2017) Influence of ultrasound on the microstructure of plant tissue. Innov Food Sci Emerg Technol 43: 117–129
44. Osae R, Zhou C, Xu B et al (2019) Nonthermal pretreatments enhances drying kinetics and quality properties of dried ginger (Zingiber officinale Roscoe) slices. J Food Process Eng 42:1–11
45. Astráin-Redín L, Alejandre M, Raso J et al (2021) Direct contact ultrasound in food processing: impact on food quality. Front Nutr 8: 1–11
46. Ashokkumar M, Sunartio D, Kentish S et al (2008) Modification of food ingredients by ultrasound to improve functionality: a preliminary study on a model system. Innov Food Sci Emerg Technol 9:155–160
47. Mawson R, Gamage M, Terefe NS et al (2011) Ultrasound in enzyme activation and inactivation. In: Feng H, Barbosa-Cánovas GV, Weiss J (eds) Ultrasound technologies for food and bioprocessing. Springer, New York, pp 369–404
48. Fernandes FAN, Gallão MI, Rodrigues S (2009) Effect of osmosis and ultrasound on pineapple cell tissue structure during dehydration. J Food Eng 90:186–190
49. Mishra V, Shyam Abrol G, Dubey N (2018) Sodium and calcium hypochlorite as postharvest disinfectants for fruits and vegetables. In: Siddiqui MW (ed) Postharvest disinfection of fruits and vegetables. Academic Press, London, pp 253–272
50. Nowacka M, Fijalkowska A, Wiktor A et al (2018) Influence of power ultrasound on the main quality properties and cell viability of osmotic dehydrated cranberries. Ultrasonics 83:33–41
51. Nowacka M, Fijalkowska A, Dadan M et al (2018) Effect of ultrasound treatment during osmotic dehydration on bioactive compounds of cranberries. Ultrasonics 83:18–25
52. Nowacka M, Tylewicz U, Laghi L et al (2014) Effect of ultrasound treatment on the water state in kiwifruit during osmotic dehydration. Food Chem 144:18–25
53. Bermúdez-Aguirre D, Mobbs T, Barbosa-Cánovas GV (2011) Ultrasound applications in food processing. In: Feng HG, Barosa-Canovas JW (eds) Ultrasound technologies

for food and bioprocessing. Springer, New York, pp 65–105

54. Bhargava N, Mor RS, Kumar K et al (2021) Advances in application of ultrasound in food processing: a review. Ultrason Sonochem 70: 105293
55. de São José JFB, de Andrade NJ, Ramos AM et al (2014) Decontamination by ultrasound application in fresh fruits and vegetables. Food Control 45:36–50
56. Huang D, Men K, Li D et al (2020) Application of ultrasound technology in the drying of food products. Ultrason Sonochem 63: 104950
57. Schössler K, Jäger H, Knorr D (2012) Novel contact ultrasound system for the accelerated freeze-drying of vegetables. Innov Food Sci Emerg Technol 16:113–120
58. Zhang J, Liu Y, Sun Z et al (2020) Drying characteristics and optimization of ultrasound-strengthened cold air drying combined with sequential far-infrared radiation drying on potato. J Food Process Preserv 44: 1–15
59. Śledź M, Nowak P, Witrowa-Rajchert D (2014) Drying of parsley leaves pre-treated by ultrasound. Adv Agric Sci Probl Issues 579: 91–99
60. Wiktor A, Sledz M, Nowacka M et al (2016) The influence of immersion and contact ultrasound treatment on selected properties of the apple tissue. Appl Acoust 103:136–142
61. Osae R, Zhou C, Alolga RN et al (2019) Effects of various nonthermal pretreatments on the physicochemical properties of dried ginger (Zingiber officinale Roscoe) slices from two geographical locations. J Food Sci 84: 2847–2858
62. Spinei M, Oroian M (2021) The influence of osmotic treatment assisted by ultrasound on the physico-chemical characteristics of blueberries (Vaccinium myrtillus L.). Ultrasonics 110: 106298
63. Rojas ML, Augusto PED (2018) Ethanol and ultrasound pre-treatments to improve infrared drying of potato slices. Innov Food Sci Emerg Technol 49:65–75
64. Fernandes FAN, Oliveira FIP, Rodrigues S (2008) Use of ultrasound for dehydration of papayas. Food Bioprocess Technol 1:339–345
65. Cichowska J, Figiel A, Stasiak-Różańska L et al (2019) Modeling of osmotic dehydration of apples in sugar alcohols and dihydroxyacetone (DHA) solutions. Foods 8:1–17
66. Nowacka M, Wedzik M (2016) Effect of ultrasound treatment on microstructure, colour and carotenoid content in fresh and dried carrot tissue. Appl Acoust 103:163–171
67. Kek SP, Chin NL, Yusof YA (2013) Direct and indirect power ultrasound assisted pre-osmotic treatments in convective drying of guava slices. Food Bioprod Process 91:495–506
68. Fijalkowska A, Nowacka M, Wiktor A et al (2016) Ultrasound as a pretreatment method to improve drying kinetics and sensory properties of dried apple. J Food Process Eng 39: 256–265
69. Nowacka M, Wiktor A, Anuszewska A et al (2019) The application of unconventional technologies as pulsed electric field, ultrasound and microwave-vacuum drying in the production of dried cranberry snacks. Ultrason Sonochem 56:1–13
70. Koubaa M, Barba FJ, Grimi N et al (2016) Recovery of colorants from red prickly pear peels and pulps enhanced by pulsed electric field and ultrasound. Innov Food Sci Emerg Technol 37:336–344
71. Sancheti SV, Gogate PR (2017) A review of engineering aspects of intensification of chemical synthesis using ultrasound. Ultrason Sonochem 36:527–543
72. Silva EK, Saldaña MDA (2020) High-intensity ultrasound-assisted recovery of cinnamyl alcohol glycosides from Rhodiola rosea roots: effect of probe diameter on the ultrasound energy performance for the extraction of bioactive compounds. Food Bioprod Process 122: 245–253
73. Szadzińska J, Łechtańska J, Kowalski SJ et al (2017) The effect of high power airborne ultrasound and microwaves on convective drying effectiveness and quality of green pepper. Ultrason Sonochem 34:531–539
74. Kowalski SJ, Pawłowski A (2015) Intensification of apple drying due to ultrasound enhancement. J Food Eng 156:1–9
75. Liu Y, Zeng Y, Hu R et al (2019) Effect of contact ultrasonic power on moisture migration during far-infrared radiation drying of kiwifruit. J Food Process Eng 42:1–10
76. Sledz M, Wiktor A, Rybak K et al (2016) The impact of ultrasound and steam blanching pre-treatments on the drying kinetics, energy consumption and selected properties of parsley leaves. Appl Acoust 103:148–156

Chapter 8

Ultrasonic Decontamination and Process Intensification

Shikha Ojha, Jefferson de Oliveira Mallia, David Spiteri, Vasilis Valdramidis, and Oliver K. Schlüter

Abstract

Ultrasound is a versatile technology and has been successfully applied in several food processes including extraction, drying, decontamination, brining, mixing and homogenization, emulsification, freezing, thawing, and cutting of foods. High-power ultrasound can induce physical and chemical changes in the biological matrices due to mechanical, cavitational, and thermal effects. This chapter outlines the method and protocols employed in application of ultrasound for food applications. In particular, operation of contact and non-contact-type ultrasound systems with a main focus on microbial decontamination and process intensification (mainly brining of meat) is described in details. Various protocols for measuring ultrasonic process-product interactions including estimation of hydrogen peroxide and oxidation products are also discussed. Furthermore, methods evaluating antimicrobial effectiveness are described in detail.

Key words Decontamination, Process intensification, Brining, Hydrogen peroxide, Sodium chloride, Antimicrobial effectiveness

1 Introduction

Low-frequency ultrasound (US) employs high-intensity waves, which induce physical and/or chemical effects on food properties, and has a strong potential for a range of food industry applications [1]. The use of US for food applications can be characterized according to frequency and power, namely, low frequency/high power (<16 to 100 kHz and power from 10 to 1000 W/cm^2) and high frequency/low power (100 kHz to 10 MHz and power <1 W/cm^2) US [2]. Low-power US is mainly used in medical diagnostics, whereas high-power US is used in applications to induce physical and chemical changes in the biological matrices due to mechanical, cavitational, and thermal effects. Implosion of cavitational bubbles, formation of microjets, microturbulence, high-velocity interparticle collisions, and perturbation in microporous particles [3] result in enhanced mass transfer and accelerated chemical reactions [4]. Formation and collapse of cavitational

Mohsen Gavahian (ed.), *Emerging Food Processing Technologies*,
Methods and Protocols in Food Science, https://doi.org/10.1007/978-1-0716-2136-3_8,

bubbles generate extremely high localized temperatures (in excess of 5000 K) and pressures (500 atm) along with formation of free radicals [5] due to sonolysis of water ($H_2O \rightarrow OH^- + H^+$, $H_2O + OH^- + H^+ \rightarrow H_2O_2 + H_2$).

2 Ultrasonic Systems, Setup, and Characterization

The use of US for decontamination has been widely demonstrated on liquid foods and foods immersed in liquid media using contact-type systems (i.e., US bath and probe-based system) [6, 7]. Non-contact ultrasonic applications have been investigated in the food industry for defoaming [8], drying [9], and inactivation of microorganisms [10].

2.1 Contact-Type Ultrasound Systems

There are different variations of contact-type US devices in the market. US bath and probe systems are the most commonly used and can be applied to research microbial inactivation and process intensification (Subheadings 4 and 5). In an US bath machine, the US transducer is responsible for converting electric energy into mechanical vibrations and is typically located under the bottom of the tank. A transducer is usually connected to a probe or a horn which may be detachable, from which US is emitted into the medium. Usually, the probe system can deliver much higher power intensities than the bath system.

Most available power US systems are fixed at usually only one frequency; however, some systems can supply dual frequencies in one machine. Most machines supply operator control options for processing time, and some offer controls for power level or amplitude and/or operation mode. For a fixed frequency wave, larger amplitude means higher power. Some machines supply different operation modes: sweep, standard, degas, and/or pulse mode. In sweep mode, the frequency varies within a defined range, causing more homogeneously distributed ultrasonic efficiency in the bath than during standard operation. In degas mode, the power is interrupted for a short period so that the ultrasonic forces do not retain the bubbles. In some probe system, pulse mode is supplied, and the US irradiation on and off periods can be set so as to reduce the average power and heat generated in the liquid to a desired level.

Typical steps and protocols for contact-type US treatment of foods are described below:

1. *Sample handling and preparation*: The samples properties can be the key factor affecting US processing and its effect. Therefore, depending on the nature of material treated, the sample handling and preparation step in detail should be recorded.

2. *Preparation of US equipment*: Before operating the equipment, it is important to conduct a thorough pre-start check on the safety features of the power connection and the equipment setup. Switch on the generator before the US process to preheat the system followed by calibration as outlined below by system manufacturer. Most US devices are equipped with an intelligent automatic frequency tuning which ensures the operation of devices at correct frequency.
3. *Operational procedure for US treatment processes*
   (a) *Power measurement*: In order to standardize the intensity level of US process due to different US devices, ultrasound intensity (UI) for the used ultrasonic horn must be calculated. The UI can be determined calorimetrically using the following equations:

$$\mathrm{UI} = \frac{4P}{\pi d^2} \tag{1}$$

where d [cm] is the diameter of the sonotrode.

In the above equation, P [W] represents the absolute ultrasonic power and can be defined as:

$$P = mC_p\left(\frac{\mathrm{d}T}{\mathrm{d}t}\right)_{x=0} \tag{2}$$

where m [g] is the mass, C_p [J/g K] is the specific heat capacity, and $\mathrm{d}T/\mathrm{d}t$ [K/s] is the ratio of change of temperature during sonication.

4. *US treatment*: Place the samples properly. Connect with external cooling circulation unit for temperature-sensitive material (if required) (*see* **Note 1**).
5. *Posttreatment analysis*: Retrieve the sample after the treatment and perform further measurements.

2.2 Airborne Ultrasound Treatment for Foods

Airborne ultrasound (AUS) utilizes non-contact transducers, which are capable of transmitting ultrasonic waves to a product using air as the coupling medium. Electrical energy is converted into mechanical vibrations in a piezoelectric transducer, which emits the ultrasonic waves through air toward the sample [11]. Ultrasonic waves may be transmitted by plate transducers by means of radiators with various shapes depending on the application as shown in Fig. 1. Plates are tuned to vibrate in flexural modes which can be circular, rectangular, or cylindrical depending on the application. The introduction of large plate surfaces has enabled an increase in the radiation resistance of air and resulting in an enhanced acoustic energy transfer between the vibrating system and the medium. The simplest plate radiator configuration is the flat-plate transducer which is characterized by a flat surface. The efficiency of this configuration

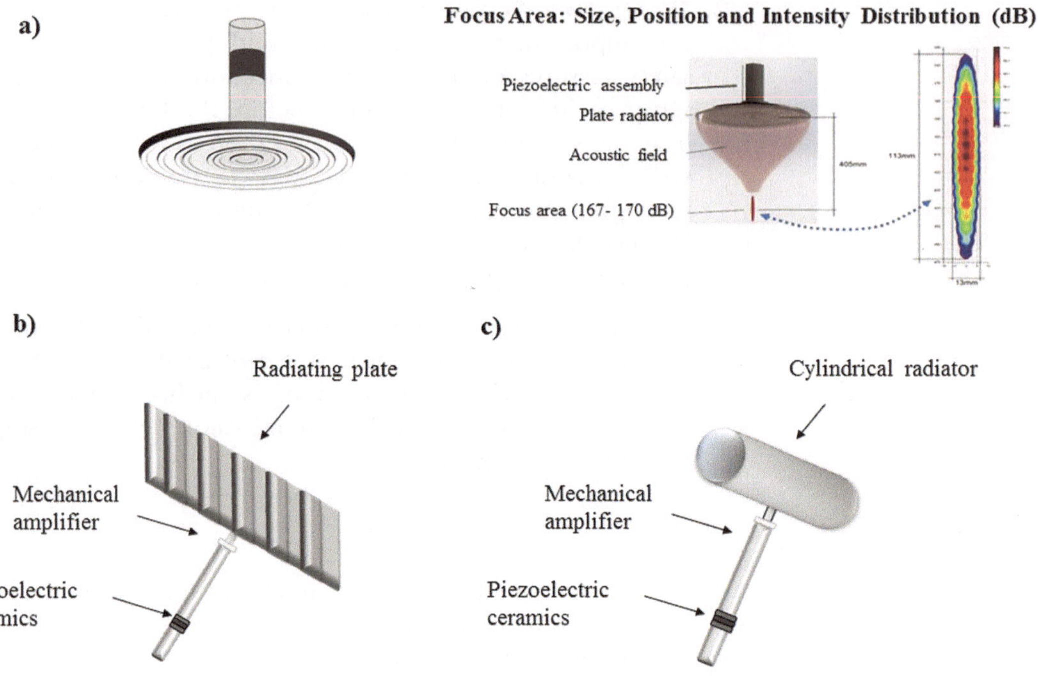

Fig. 1 Airborne acoustic transducers: (**a**) circular stepped-plate; (**b**) rectangular stepped-plate; (**c**) cylindrical plate (from [11])

is limited in the far field due to generation of lateral lobes in the acoustic pattern. The design of transducers has significantly improved recently enabling tailoring of the acoustic fields according to specific operational requirements. Highly directive coherent fields similar to those produced by a piston have been achieved by the addition of steps to the radiator surfaces, while steps and grooves have been shown to be capable of generating highly focused fields.

The efficiency and effectiveness of AUS processing is strongly dependent on the physiochemical properties of the food products used, sample preparation prior to processing, processing parameters, and the design of equipment. A typical system consists of an electronic power generator with a dynamic resonance controller, a power amplifier, a high impedance matching box, and a circular stepped-plate transducer. This type of transducer produces a highly focused acoustic field [12].

Typical steps and protocols for airborne US treatment of foods are described below:

1. *Sample handling and preparation prior to AUS treatment*: The samples properties can be the key factor affecting AUS processing and its effect; therefore, the entire sample handling and preparation steps should be recorded in detail.

2. *Preparation of AUS equipment*: Before operating the equipment, it is important to conduct a thorough pre-start check on the safety features of the equipment setups, as outlined below.
   (a) Mount the equipment safely by positioning it on a flat surface.
   (b) Ensure that the space where the equipment is located is well ventilated and there is no obstruction around the equipment.
   (c) Check the main power supply voltage and frequency to be connected to the equipment. Be aware of the location electrical safety symbols and electrical hazard warning signs of the equipment that can cause serious personal injury or equipment damage if ignored.
   (d) Check all the connections, especially the wire that connects the generator and the transducer (*see* **Note 2**).
   (e) Adjust the height of the system to fit the samples (*see* **Note 3**).
3. *Operational procedure for AUS treatment process*
   (a) Plug in the power meter.
   (b) Switch on the generator 20 min before AUS process to preheat the system.
   (c) Load the samples properly in the range of tailored hood (*see* **Note 4**).
   (d) Put on ear defenders and switch on the amplifier.
   (e) Retrieve the sample after treatment and do measurements.
4. *Monitoring output voltage and current*: Once the process is done, the output voltage and current can be obtained by the power meter.
5. *Evaluating the effect of AUS treatment*: It is crucial to determine whether the applied AUS treatment affect the cell microstructure and porosity. As a result of AUS treatment, the food sample is expected to experience subtle damage. Cárcel, García-Pérez [13] and Ozuna, Álvarez-Arenas [14] reported airborne US-assisted convective drying could generate microchannels in their apple samples, caused by "sponge effect."
   (a) Direct measurement of US intensity.
   (b) Estimation of the qualities of the samples: Freeze dry and store the representative specimens before and after AUS treatment. Follow the specific protocols for physiochemical analysis.
   (c) Microstructural analysis (suitable for solid foods).
      - Prepare the representative specimens before and after AUS treatment.

- Employ microscopic imaging techniques such as light, fluorescent, electron (SEM, cryo-SEM, TEM), or optical coherence tomography technique to identify the structural changes.

(d) Total viable count measurement: Prepare the representative specimens before and after AUS treatment, and dilute with maximum recovery diluent (MRD). Use pour plate or spread plate method to inoculate the samples on plate count agar (PCA) (*see* Subheading 4.1).

3 Ultrasonic-Induced Chemical Reactions

Ultrasonic cavitation causes the formation of gaseous nuclei within a liquid, with the subsequent disintegration of these bubbles [15–17]. Cavitation can be classed as stable and transient cavitation. In stable cavitation, these bubbles can persist for many acoustic cycles and are formed at low ultrasonic intensities (1–3 W/cm^2). In contrast, bubbles produced during transient cavitation will only survive a few cycles before violently collapsing due to higher intensities (>10 W/cm^2). This violent collapse results in the generation of high pressures and temperatures (>50 MPa and >3000 K) at the gas-liquid interface, which allows for the homolytic fission of molecular bonds. Most notably in aqueous environments is the breakdown of OH bonds in water molecules to form hydrogen (H•) and hydroxyl (HO•) free radicals.

The assessment of free radicals has attracted a lot of interest from the medical fields as these are involved in various biological process [18]. Their generation post-US treatments within in vivo and in vitro environments has been extensively studied [19, 20]. This allowed for the development of a variety of techniques to investigate their interactions [21]. The effect of free radicals within foods has been studied and focus on endpoint assessments that investigate changes in major nutritional compounds such as lipids, protein/amino acids, and vitamins [22].

The formation of free radicals during US treatments of food products can have positive impacts on their resultant quality as it acts as a decontamination process, thus improving their preservation [23, 24]. Still, the excessive formation of free radicals could result in adverse quality effects [25]. Here the formation of reactive oxygen species (ROS) would cause chemical changes in various compounds within these substances, potentially diminishing their nutritional and sensory properties. During US treatments, the formation of hydrogen peroxide (H_2O_2) as a long-lived ROS is well known and involved in the previous indicated processes [26]. Assay procedures that can be used to quantify H_2O_2 generated using titanium sulfate and substance oxidation using ferrous bound xylenol orange post-US treatment are outlined here.

3.1 Procedure to Determine Hydrogen Peroxide Generated Using Titanium Sulfate Assay

This assay involves the preparation of titanium sulfate, $TiSO_4$ reagent [27], that specifically reacts with H_2O_2 to form pertitanic acid (H_2TiO_4) that produces a yellow color. The generation of pertitanic acid can be assessed by measuring its absorbance at 410 nm. The following procedure is designed to use 1 cm quartz cuvettes with the appropriate spectrometer but can be adapted for 96-well plate readers.

1. Preparation of $TiSO_4$ (Eisenberg) reagent.
   (a) In a fume hood, prepare a sand bath and ensure that hot plate temperature is set at 150 °C.
   (b) With a 500 mL round bottom flask glass volumetric flask, add 200 mL of 96% sulfuric acid (H_2SO_4, CAS no.: 7664-93-9) and transfer 2 g of anhydrous TiO_2 (CAS no.: 13463-67-7). Leave the reaction flask for about 15 h at 150 °C.
   (c) Once complete, leave the mixture to cool to room temperature. Store the mixture in at room temperature and is stable for at least 6 months.
   (d) In a fume hood, prepare the $TiSO_4$ reagent by diluting 70 mL of the previous mixture in 210 mL of distilled water (H_2O).
   (e) Using a grade 4 (pore size $\leq$16–40 μm) or higher grade sintered glass funnel and a vacuum filtration setup, filter the diluted solution. The resultant filtrate is the assay detection reagent. This can be stored at room temperature for later use and is stable for at least 6 months.
2. Preparation of the H_2O_2 standard curve is done by diluting 30% (wt.) H_2O_2 (CAS no.: 7722-84-1) to make a stock solution of 100 ppm. This stock solution is then used to prepare 0.0, 5.0, 10.0, 15.0, 20.0, and 25.0 ppm standards. The standard curve can be adapted to represent better the test range being investigated.
3. Standard and sample analysis is done as follows:
   (a) Transfer 5 mL of standard or sample and add 500 μL of reagent.
   (b) Briefly vortex the resultant solution and incubate for 5 min at room temperature.
   (c) Measure absorbance at 410 nm for each solution in duplicate.
   (d) Using the standard curve, determine the concentration of H_2O_2 in each sample. The limits of detection and quantification should be determined as well.

3.2 Procedure to Determine Substance Oxidation Using Ferrous Oxidation in Xylenol Orange (FOX) Assay

This assay involves the oxidation of ferrous (Fe^{2+}) to ferric (Fe^{3+}) ions that bind to xylenol orange molecules, generating a ferric complex that produces an orange color [28]. The formation of the resultant complex can be assessed by measuring its absorbance at 560 nm. The following procedure is designed to use 1 cm quartz cuvettes with the appropriate spectrometer but can be adapted for 96-well plate readers.

1. Preparation of 2.5 M H_2SO_4 is carried out in a fume hood. This is done by diluting 96% H_2SO_4 (CAS no.: 7664-93-9) in distilled H_2O to obtain the required concentration.
2. To prepare 25 mM ammonium ferrous(II) sulfate in 2.5 M H_2SO_4 (reagent A):
   (a) In a 100 mL glass volumetric flask, transfer about 50 mL of 2.5 M H_2SO_4.
   (b) Weigh 0.9804 g of ammonium ferrous(II) sulfate hexahydrate (CAS no.: 7783-85-9) and transfer this to the flask.
   (c) Top up with 2.5 M H_2SO_4 until the 100 mL mark and mix until all of the solute has dissolved.
3. To prepare the 100 mM sorbitol and 125 μM xylenol orange solution (reagent B):
   (a) In a 250 mL amber glass volumetric flask, transfer 100 mL of distilled H_2O.
   (b) Weigh 4.5545 g of D-sorbitol and transfer this to the flask.
   (c) Weigh 0.0224 g of xylenol orange and transfer this to the flask.
   (d) Top up with distilled H_2O until the 250 mL mark.
4. Reagents A and B must be stored at 4 °C and used within 1 month.
5. To prepare the assay detection reagent,
   (a) Mix 1 volume of reagent A with 100 volumes of reagent B by transferring 100 mL of reagent B and adding 1 mL of reagent A.
   (b) Mix the resultant solution (*see* **Note 5**).
6. Preparation of the H_2O_2 standard curve is done by diluting 30% (wt.) H_2O_2 (CAS no.: 7722-84-1) to make a stock solution of 100 ppm. This stock solution is then used to prepare 0.0, 0.7, 1.4, 2.1, 2.8, and 3.5 ppm standards (*see* **Note 6**).
7. Standard and sample analysis is done as follows:
   (a) Transfer 4 mL of assay detection reagent and add 100 μL of sample or standard.
   (b) Briefly vortex the resultant solution and incubate for 20 min at room temperature.

(c) Measure absorbance at 560 nm for each solution in duplicate.

(d) Using the standard curve, determine the concentration of peroxide in each sample. The limits of detection and quantification should be determined as well.

3.3 General Considerations for $TiSO_4$ and FOX Assays

These assays offer a relatively straightforward endpoint analysis to assess ROS formation post-US treatments. The increases in absorption for these assays at their specified wavelengths can be correlated to increases in ROS generation after the various treatments. Still interferences have to be considered when analyzing food products post-US treatment [28–30]. Spiked samples should be assessed to determine antioxidant, reducing and interfering effects. Compounds released during US treatment might have absorb at the same wavelength and bias the measurements. Therefore, absorbance measures of treated samples without the addition of assay reagents should be performed and subtracted from the final absorbance. Particulate matter released during these treatments is another concern; centrifugation and filtration steps would need to be included to remove these interferences. When multicomponent mixtures are being assessed, interactions must be specifically attributed to the specific parts of these mixtures. Model solutions should therefore be included in these investigations [24, 31, 32]. These could consist of the major components or specific components of the test substance. The exclusion and inclusion of components would provide a better understanding of the ROS interactions occurring during US treatments.

4 Procedures for Ultrasonic Decontamination

In recent years, the use of ultrasonic technology in water and wastewater treatment, as well as ecological remediation, including sanitization, has received a lot of attention [33, 34]. US produces elastic vibrations and waves with a frequency of more than 15–20 kHz. At low intensities and for short periods, US can induce microbial function and proliferation, but at higher intensities, it destroys and inactivates microorganisms. US with a frequency of 20–100 kHz and a sound intensity of 10–1000 W/cm^2 can be used to disinfect water for a long time [35].

Sonication's disinfection potential in liquids is due to the phenomenon of acoustic cavitation, which is the formation and collapse of microbubbles in milliseconds, causing extreme temperature and pressure gradients [36, 37]. When these microbubbles burst, the surrounding atmosphere is subjected to extremely high temperatures and pressures. Extremely reactive radicals have been observed in these conditions. US can inactivate bacteria and deagglomerate bacterial clusters due to the physical, mechanical, and chemical effects of acoustic cavitation [38].

To evaluate the antimicrobial effectiveness and mechanism of action of US systems, a variety of protocols can be used.

- Viable counts of microorganisms.
- Viable counts of different mutant cells.
- Transcriptomic study.
- Sequencing of the whole genome.
- Other methods, such as flow cytometry, comet assay, and so on.

4.1 Viable Counts of Microorganisms

Food safety quality standards are mainly oriented toward the intended use and are outlined in international guidelines and legal requirements. Many environmental factors such as temperature fluctuations, air, and light can spoil food and water while also providing a breeding ground for bacteria. Bacterial pathogens, including indigenous aquatic bacteria, viral pathogens, and protozoan parasites, are all present in the environment. US has been related to oxidative stress in *Saccharomyces cerevisiae* [39], *Francisella tularensis* [40], *Escherichia coli* [41], and *Listeria monocytogenes* [42].

Serial plate counting using conventional microbiology methods and species-specific media can be a quick way to detect microorganism reduction after US treatments.

1. Aliquots (1 mL) are collected from US treated solutions.
2. Serial dilutions of samples at different times are performed, and the appropriate dilution is spread on tryptic soy agar (TSA) plates for assessing natural microflora.
3. Samples are incubated at 30 °C for 48 h.
4. Low microbial population counts are assessed by plating 1 mL of the sample over three TSA plates according to ISO 7218: 2007.
5. Microbial counts are expressed as log CFU/mL (*see* **Note** 7).

4.2 Viable Counts of Different Mutant Cells

Microbial mutants may be used to study the antimicrobial mechanism of action of US. Mutants with particular genes knocked out, for example, can aid in determining the effect of US on different metabolic and biological factors within the cell. The role of the gene can determine the effect of the treatment on the cell, so in the case of knocking out certain genes, the effect on the cell can be determined by the treatment. After 3 min of continuous US treatments (200 W, frequency of 26 kHz), literature indicates that the mutants oxyR were more resistant to the treatment, while dnaK was almost as susceptible as the wild form [41]. According to Patil, Valdramidis [43], the soxR, soxS, oxyR, rpoS, and dnaK genes play a significant role in the protection against reactive oxygen radicals. One of the phenomena induced by cavitation is the formation of radicals H• and OH•, as well as H_2O_2 [44], which is known to cause oxidative stress in bacteria [44]. The dnaK protein, in particular, is required for high-temperature growth and is involved in the regulation of the heat shock response. The heat shock response is a cellular response to a variety of stresses, including heat, ethanol, oxidants, and DNA-damaging agents, abnormal protein growth, viral infections, and nutrient deprivation [45]. The oxyR regulates the expression of the oxyR regulon, which consists of many genes. The oxyR protein, which is formed constitutively, is oxidized by H_2O_2. oxyR binds to target gene promoters in its oxidized form and activates transcription by enabling protein-protein interaction with RNA polymerase. In the cell's defense, oxyR-activated genes play both direct and indirect antioxidant functions, such as catalase's removal of H_2O_2 and the Dps protein's protection of DNA from oxidative attack [46].

Similar protocols of viable count assessments with all the aforementioned mutants can be performed following the previously described protocol.

4.3 Transcriptomic Analysis

Transcriptomics has recently been used to investigate the impact of various microbial stresses. Wecke and Macher coined the word "omics era" to describe this time in 2011. Our understanding of cell resistance mechanisms and/or regulatory networks that organize bacterial stress responses has improved as a result of studying gene expression through looking at RNA transcripts present in cells [47]. Several studies of the transcriptome of *E. coli* during such stress responses have added to our understanding of stress mechanisms (e.g., Chueca, Pagán [47]; Harcum and Haddadin [48]; King, Lucchini [49]; Royce, Boggess [50]; Yung, Grasso [51]; Zheng, Wang [52]; Li, Zhang [53]). Many antibiotics' modes of action, as well as mechanisms of bacterial adaptation and inactivation by heat or high hydrostatic pressure, have been identified in these studies [54].

In addition, transcriptional profiling has shown the activation of general stress responses and proteins following particular methods of treatment due to specific cross-resistance phenomena [47, 55, 56]. As a result, the first heat shock response in *E. coli* was identified in 1978 by Harcum and Haddadin [48]. They discovered that 20 proteins were heat sensitive and were influenced by transcription levels. This discovery ignited a flurry of others, revealing that certain heat shock proteins facilitate protein folding while others function as proteases [48]. Further research using RNA-Seq methods, which can capture the global transcriptional response during specific conditions in any organism, could allow for simultaneous analysis of all regions within the genome, as opposed to other methods like RT-PCR, which are still limited to analyzing specific and recognized genomic regions.

The aim of conducting transcriptomic analysis is to determine which genes are turned on or off as a result of a treatment. It is typically vital to detect such differences right after treatment, but in most situations, in order to operate with viable cells, the bacteria may have reversed the differences that the US would have caused. Working with viable cells, on the other hand, means that any changes that occur inside the cell can result in long-term changes. Once the cells are isolated, RNA extraction can be carried out by a number of ready-made kits.

1. Check the QC of the extracted RNA by using bioanalyzers. If sufficient quality is achieved, the RNA library is prepared by attaching oligo-dT probes. This allows the poly-A RNA to be selected that produces the mRNA.
2. Break down the fragments with high temperatures to 200 bp log fragments, which are used to prepare cDNA libraries. At this point, the libraries are read on an analyzer to perform data analysis, called bioinformatics.
3. A number of software packages are available for such analysis. These software packages carry out a number of algorithms in order to compare the libraries extracted from the bacteria under study with reference genomes, found on databases. This type of analysis is carried out by preparing contigs from the samples studied.
4. Reference these contigs and annotate against the reference genomes downloaded from an online database.
5. Once these are compared, the packages would be able to export all the RNA genes expressed. The results can also be plotted visually by running volcano and PCA plots.

Venn diagrams and heat maps also visually express the genes obtained and easily show patterns between different treatments. These patterns would allow the researcher to search for specific genes in Gene Ontology databases and obtain specific pathways and mechanisms (KEGG).

4.4 Whole Genome Sequencing

Direct measurements of mutation frequencies are now possible thanks to whole genome sequencing [56]. This is particularly true in the case of *E. coli*, where 12 populations have been reported to spread independently for over 40,000 years [57, 58]. Microorganisms have been used to research adaptive evolution mechanisms because of their short generation times, large population size, accuracy of experiment results, and ability to maintain ancestor strains [59].

Pavlov and Ehrenberg proposed a model in which bacteria manipulate gene expression to rapidly respond to environmental changes in 2013 [60]. Their research revealed that bacteria can direct their resources toward cell growth by growing pathways that cause other sources of energy to improve metabolism rather than multiplying when faced with environmental stress [61, 62]. Ferenci [63] investigated this further and found a highly unbalanced relationship between stress and mutations in *E. coli*. As a result, cells in certain physiological conditions or carrying out specific mutations may be able to withstand non-thermal physical treatments [62, 64]. Studying the mechanisms of microorganism responses to US can aid in understanding the potential risks and preventing any potential safety accidents, such as induced resistance, uncontrolled mutations, and enhanced recovery. More research is needed to better understand how microorganisms react to US stress, such as oxidative stress systems, cell repair, and resistance regulation [62]. Furthermore, many scientists have been perplexed by the maintenance of genetic systems that seem to have no selective advantages. "Cryptic" genes, which have no role in wild-type organisms and require mutational activation for expression, are part of such a genetic system [65].

NGS analysis is carried out by using beads washed with extraction buffer and containing lysozyme and RNase A.

1. Once purification of the genomic DNA is completed, the libraries are prepared using manufacturer-specific library preparation kits.
2. This will allow the genome libraries to be sequenced by the analyzer.
3. The data obtained can then be analyzed by running BBDuk and preparing contigs.
4. These contigs are compared with the reference genomes, in order to find SNPs in the mapped data. These SNPs will give rise to mutations within the genome after treatment.

4.5 Other Techniques

Further studies in the gene expression may also validate or enhance the work performed in such research. Transcriptomic analysis is a very strong tool in the identification of gene expression; however such expression will easily revert back once the US stress conditions are not present anymore. For this reason, RT-qPCR tests can be carried out to confirm the significant transcriptional difference between genes in both treatment samples, using different strains. Specific genes can be selected to calculate fold changes between treatment and control samples as described by Gallup and Ackermann [65].

5 Ultrasonic Process Intensification

US process intensification makes the reactions and the production of materials more efficient by improving rates and the output yields. Ultrasonic can be used in a range of processes including diffusion, homogenization, dispersion and wet milling, emulsification, extraction, lysis, and sonochemical reactions. For instance, brining of the meat is one of the major technologies used in processed meat manufacture, as it enhances shelf life, flavor, juiciness, and tenderness of the products. However, the migration of NaCl from the brine to the meat matrix is normally quite slow [66, 67]. Power US treatment can modify cell membranes through cavitation, which can help with curing, marinating, drying, and tenderizing the meat tissue, therefore helping in the enhancement of food quality and safety profile of the products [68]. US can help to reduce brining time without significant negative changes in other characteristics of the meat such as changes in quality (color, texture, cook loss, expressible moisture), sensory attributes, oxidative stability, and microbial load [69, 70]. Moreover, power US, acting on meat texture and providing a better distribution of salt in the meat matrix, could be helpful in the development of reduced salt meat formulations [71].

5.1 Procedure for US-Assisted Brining

A typical procedure for brining of meat includes the following steps:

1. *Sample preparation*: Muscles are generally stored at 4 °C prior to being processed. The pH of all the muscles should be 5.4 ± 0.4 when recorded by direct insertion of a pH electrode along the length of the muscle. Before curing, the connective tissue is carefully trimmed from the surface of the meat.
2. *Brine preparation*: A desired concentration of salt solution is prepared (e.g., 15% (w/w) NaCl). Additional curing ingredients (sugar, preservatives, spices, etc.) can be added for certain kind of products.
3. *US system setup*: Brining of meat can be done either with a probe, bath system, or combination. Figure 2 presents research on US enhancement of salt diffusion in meat using both an US bath and a probe system. Meat is cured in brine solution in the US tank, with US irradiated from the bottom of the tank or from the probe from the top of the tank.
4. *Temperature control*: The temperature of the brine solution is maintained using an external cooling system, consisting of a refrigerated circulator, a heat exchanger, and a variable flow rate pump. Maintaining the temperature of brine is very important to minimize the thermal effect of US treatment.

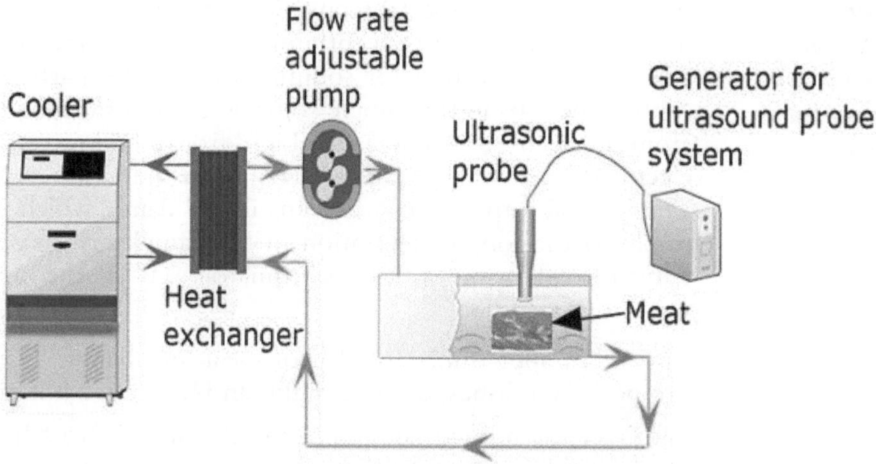

Fig. 2 Schematic diagram of experimental setup for brining of meat

5. *Brining*: Add desired brine solution in ultrasonic bath system at a desired frequency and treatment temperature. The ratio between the meat and the brine (e.g. 1:40 or higher) should be chosen to ensure no variations in the salt concentration of the brine between treatments.
6. *Further analyses*: After curing for sufficient interval, samples are rinsed with distilled water and stored for further analyses.

5.2 Method for Determination of Chloride Ion Concentration by Titration (Mohr's Method)

The Mohr titration is a direct titration method to quantitate chloride ions, to then calculate salt content. The method uses chromate ions as an indicator in the titration of chloride ions with a silver nitrate standard solution. After all the chloride has been precipitated as white silver chloride, the first excess of titrant results in the formation of a silver chromate precipitate, which signals the endpoint.

1. Prepare 0.1 M silver nitrate solution and standardize against primary standard KCl solution.
2. Weigh about 10 g of the sample and transfer it quantitatively to a conical flask, add 100 mL of hot water, and boil the content for 15 min with repeated shaking (*see* **Note 8**).
3. Filter the solution through glass wool. Transfer 50 mL of each solution to 250 mL Erlenmeyer flasks.
4. Add 1 mL of potassium chromate indicator to each 50 mL of filtrate.
5. Titrate the solution with standardized ca. 0.1 M $AgNO_3$, to the first visible pale red-brown color that persists for 30 s. Record the volume of titrant used.
6. Calculate the chloride content and the sodium chloride content in terms of percent and weight/volume.

5.3 Method for Determination of Sodium Content with Atomic Absorption Spectroscopy

Atomic absorption spectroscopy (AAS) method can be used for the determination of the concentration of specific minerals in foods. The limit of detection for sodium is up to 0.3 parts per billion (ppb). The ions in the sample solution are transformed to neutral atoms in an air/acetylene flame. Light from a hollow cathode or an electrodeless discharge lamp (EDL) is passed through the flame. The light absorption of the atoms in the flame, which is proportional to the ion concentration in the sample, is measured by a detector following a monochromator set at the appropriate wavelength.

1. Weigh approximately 10 g of meat and blend. Transfer into porcelain dishes, dry overnight, and burn on a hot plate.
2. Place the burned samples in a muffle furnace to ash at 525 °C for approximately 8–10 h.
3. Cool the ashes in desiccator and dissolve ash in few drops of hydrochloric acid (HCl). If necessary, filter the samples and make appropriate dilution of the samples (e.g., dilute to a volume of 50 mL) in a volumetric flask.
4. Sodium standard solutions were prepared using known concentrations of sodium chloride (NaCl) and calibration curves to correlate the relative absorbance.
5. Spike each standard, control, and sample 9:10 with 0.5% lanthanum chloride solution (one part of solution and nine parts standard, control, or sample) (*see* **Note 9**).
6. Follow manufacturer's instructions for startup, use, and shutdown of the AAS. Generally, the instrument should be recalibrated after every 20–30 samples. A control solution should also be run after each calibration.

6 Notes

1. Put on ear defenders and start the treatment at desired amplitude or power value.
2. Be careful with the wire that connects the generator and transducer because it is fragile.
3. The transducer and plate are very heavy. When operating the system, it is better to have two operators to adjust the system.
4. AUS is normally studied with the combination of other equipment such as fluidized bed dryer, plasma system, etc.; it is important to do trials to figure out the suitable sample conditions before the operation.
5. This volume will be enough to measure 12 standard and/or samples. It must be always prepared fresh on the day of analysis.

If more measurements must be carried out, prepare as necessary.

6. The standard curve can be adapted to represent better the test range being investigated.
7. Only counts between 30 and 300 CFU/mL are considered accurate measurements as they are easily read.
8. The salt can also be extracted from the food sample by means of ashing. Weigh the samples (10 g) into porcelain dishes, dry overnight, and place on a hot plate until completely burnt. Place the dishes containing the burned samples in the muffle furnace at 525 °C for approximately 8–10 h.
9. Lanthanum solution is added to prevent chemical and ionization interference.

References

1. Kentish S, Feng H (2014) Applications of power ultrasound in food processing. Annu Rev Food Sci Technol 5(1):263–284
2. Soria AC, Villamiel M (2010) Effect of ultrasound on the technological properties and bioactivity of food: a review. Trends Food Sci Technol 21(7):323–331
3. Shirsath S, Sonawane S, Gogate P (2012) Intensification of extraction of natural products using ultrasonic irradiations—a review of current status. Chem Eng Process 53:10–13
4. Kadam SU, Tiwari BK, O'Donnell CP (2013) Application of novel extraction technologies for bioactives from marine algae. J Agric Food Chem 61(20):4667–4675
5. Rokhina EV, Lens P, Virkutyte J (2009) Low-frequency ultrasound in biotechnology: state of the art. Trends Biotechnol 27(5): 298–306
6. Piyasena P, Mohareb E, McKellar R (2003) Inactivation of microbes using ultrasound: a review. Int J Food Microbiol 87(3):207–216
7. Tiwari B, Mason T (2012) Ultrasound processing of fluid foods. In: Novel thermal and non-thermal technologies for fluid foods. Elsevier, Amsterdam, pp 135–165
8. Rodríguez G et al (2010) Experimental study of defoaming by air-borne power ultrasonic technology. Phys Procedia 3(1):135–139
9. Zhu X et al (2021) Applications of ultrasound to enhance fluidized bed drying of Ascophyllum nodosum: drying kinetics and product quality assessment. Ultrason Sonochem 70: 105298
10. Pisano MA, Boucher RM, Alcamo IE (1966) Sterilizing effects of high-intensity airborne sonic and ultrasonic waves. Appl Microbiol 14(5):732–736
11. Charoux CM et al (2017) Applications of airborne ultrasonic technology in the food industry. J Food Eng 208:28–36
12. Gallego-Juárez JA, Riera E (2011) Technologies and applications of airborne power ultrasound in food processing. In: Ultrasound technologies for food and bioprocessing. Springer, Berlin, pp 617–641
13. Cárcel J et al (2012) Food process innovation through new technologies: use of ultrasound. J Food Eng 110(2):200–207
14. Ozuna C et al (2014) Influence of material structure on air-borne ultrasonic application in drying. Ultrason Sonochem 21(3): 1235–1243
15. Wu TY et al (2013) Theory and fundamentals of ultrasound. In: Wu TY et al (eds) Advances in ultrasound technology for environmental remediation. Springer, Dordrecht, pp 5–12
16. Riesz P, Berdahl D, Christman CL (1985) Free radical generation by ultrasound in aqueous and nonaqueous solutions. Environ Health Perspect 64:233–252
17. Santos HM, Lodeiro C, Capelo-Martínez J-L (2008) The power of ultrasound. Ultrasound Chem:1–16
18. Phaniendra A, Jestadi DB, Periyasamy L (2015) Free radicals: properties, sources, targets, and their implication in various diseases. Indian J Clin Biochem 30(1):11–26
19. Ogawa R et al (2016) Bioeffects of ultrasound and its therapeutic application. In: Handbook of ultrasonics and sonochemistry. Springer Singapore, Singapore, pp 1049–1074

20. Tang J, Guha C, Tome WA (2015) Biological effects induced by non-thermal ultrasound and implications for cancer therapy: a review of the current literature. Technol Cancer Res Treat 14(2):221–235
21. Zhang Y, Dai M, Yuan Z (2018) Methods for the detection of reactive oxygen species. Anal Methods 10(38):4625–4638
22. Choe E, Min DB (2006) Chemistry and reactions of reactive oxygen species in foods. Crit Rev Food Sci Nutr 46(1):1–22
23. Deng Y et al (2018) Influence of ultrasound assisted thermal processing on the physicochemical and sensorial properties of beer. Ultrason Sonochem 40(Pt A):166–173
24. Zhang QA et al (2015) Free radical generation induced by ultrasound in red wine and model wine: an EPR spin-trapping study. Ultrason Sonochem 27:96–101
25. Pingret D et al (2012) Degradation of edible oil during food processing by ultrasound: electron paramagnetic resonance, physicochemical, and sensory appreciation. J Agric Food Chem 60(31):7761–7768
26. Ziembowicz S, Kida M, Koszelnik P (2017) Sonochemical formation of hydrogen peroxide. In: Proceedings, vol 2(5), p 188
27. Eisenberg G (1943) Colorimetric determination of hydrogen peroxide. Ind Eng Chem Anal Ed 15(5):327–328
28. Bou R et al (2008) Determination of hydroperoxides in foods and biological samples by the ferrous oxidation-xylenol orange method: a review of the factors that influence the method's performance. Anal Biochem 377(1):1–15
29. Satterfield CN, Bonnell AH (1955) Interferences in titanium sulfate method for hydrogen peroxide. Anal Chem 27(7):1174–1175
30. Amin VM, Olson NF (1967) Spectrophotometric determination of hydrogen peroxide in milk. J Dairy Sci 50(4):461–464
31. Ashokkumar M et al (2008) Modification of food ingredients by ultrasound to improve functionality: a preliminary study on a model system. Innov Food Sci Emerg Technol 9(2): 155–160
32. Kiani H et al (2011) Ultrasound assisted nucleation of some liquid and solid model foods during freezing. Food Res Int 44(9): 2915–2921
33. Cesaro A, Belgiorno V (2016) Removal of endocrine disruptors from urban wastewater by advanced oxidation processes (AOPs): a review. Open Biotechnol J 10(1):151–172
34. Han C et al (2013) Green nanotechnology: development of nanomaterials for environmental and energy applications. In: Sustainable nanotechnology and the environment: advances and achievements. ACS Publications, Washington, DC, pp 201–229
35. Vasilyak L (2010) Ultrasound application in systems for the disinfection of water. Surf Eng Appl Electrochem 46(5):489–493
36. Drakopoulou S et al (2009) Ultrasound-induced inactivation of gram-negative and gram-positive bacteria in secondary treated municipal wastewater. Ultrason Sonochem 16(5):629–634
37. Sango DM et al (2014) Assisted ultrasound applications for the production of safe foods. J Appl Microbiol 116(5):1067–1083
38. Antoniadis A et al (2007) Sonochemical disinfection of municipal wastewater. J Hazard Mater 146(3):492–495
39. Coleman ST et al (2001) Expression of a glutamate decarboxylase homologue is required for normal oxidative stress tolerance in Saccharomyces cerevisiae. J Biol Chem 276(1):244–250
40. Ramond E et al (2014) Glutamate utilization couples oxidative stress defense and the tricarboxylic acid cycle in Francisella phagosomal escape. PLoS Pathog 10(1):e1003893
41. Spiteri D et al (2017) Ultrasound processing of liquid system(s) and its antimicrobial mechanism of action. Lett Appl Microbiol 65(4): 313–318
42. Boura M, Brensone D, Karatzas KA (2020) A novel role for the glutamate decarboxylase system in Listeria monocytogenes; protection against oxidative stress. Food Microbiol 85: 103284
43. Patil S et al (2011) Assessing the microbial oxidative stress mechanism of ozone treatment through the responses of Escherichia coli mutants. J Appl Microbiol 111(1):136–144
44. Joyce E et al (2003) The development and evaluation of electrolysis in conjunction with power ultrasound for the disinfection of bacterial suspensions. Ultrason Sonochem 10(4–5): 231–234
45. Bukau B, Walker GC (1989) Cellular defects caused by deletion of the Escherichia coli dnaK gene indicate roles for heat shock protein in normal metabolism. J Bacteriol 171(5): 2337–2346
46. Pomposiello PJ, Demple B (2001) Redox-operated genetic switches: the SoxR and OxyR transcription factors. Trends Biotechnol 19(3):109–114
47. Chueca B, Pagán R, García-Gonzalo D (2015) Transcriptomic analysis of Escherichia coli MG1655 cells exposed to pulsed electric fields. Innovative Food Sci Emerg Technol 29:78–86

48. Harcum SW, Haddadin FT (2006) Global transcriptome response of recombinant Escherichia coli to heat-shock and dual heat-shock recombinant protein induction. J Ind Microbiol Biotechnol 33(10):801–814
49. King T et al (2010) Transcriptomic analysis of Escherichia coli O157: H7 and K-12 cultures exposed to inorganic and organic acids in stationary phase reveals acidulant- and strain-specific acid tolerance responses. Appl Environ Microbiol 76(19):6514–6528
50. Royce LA et al (2014) Transcriptomic analysis of carboxylic acid challenge in Escherichia coli: beyond membrane damage. PLoS One 9(2): e89580
51. Yung PY et al (2016) Global transcriptomic responses of Escherichia coli K-12 to volatile organic compounds. Sci Rep 6(1):1–15
52. Zheng M et al (2001) DNA microarray-mediated transcriptional profiling of the Escherichia coli response to hydrogen peroxide. J Bacteriol 183(15):4562–4570
53. Li Y et al (2019) Systematic identification and validation of the reference genes from 60 RNA-Seq libraries in the scallop Mizuhopecten yessoensis. BMC Genomics 20(1):1–12
54. Wecke T, Mascher T (2011) Antibiotic research in the age of omics: from expression profiles to interspecies communication. J Antimicrob Chemother 66(12):2689–2704
55. Carruthers MD, Minion C (2009) Transcriptome analysis of Escherichia coli O157: H7 EDL933 during heat shock. FEMS Microbiol Lett 295(1):96–102
56. Shin J-H et al (2010) σB-dependent protein induction in Listeria monocytogenes during vancomycin stress. FEMS Microbiol Lett 308(1):94–100
57. Batt CA (2014) ESCHERICHIA COLI | Escherichia coli. In: Batt CA, Tortorello ML (eds) Encyclopedia of food microbiology (second edition). Academic Press, Oxford, pp 688–694
58. Philippe N et al (2007) Evolution of global regulatory networks during a long-term experiment with Escherichia coli. BioEssays 29(9): 846–860
59. Elena SF, Lenski RE (2003) Evolution experiments with microorganisms: the dynamics and genetic bases of adaptation. Nat Rev Genet 4(6):457–469
60. Pavlov MY, Ehrenberg M (2013) Optimal control of gene expression for fast proteome adaptation to environmental change. Proc Natl Acad Sci 110(51):20527–20532
61. Berney M, Weilenmann H-U, Egli T (2007) Adaptation to UVA radiation of E. coli growing in continuous culture. J Photochem Photobiol B Biol 86(2):149–159
62. Wu D et al (2020) Microbial response to some nonthermal physical technologies. Trends Food Sci Technol 95:107–117
63. Ferenci T (2019) Irregularities in genetic variation and mutation rates with environmental stresses. Environ Microbiol 21(11): 3979–3988
64. Li J et al (2017) Synergetic effects of ultrasound and slightly acidic electrolyzed water against Staphylococcus aureus evaluated by flow cytometry and electron microscopy. Ultrason Sonochem 38:711–719
65. Gallup JM, Ackermann MR (2006) Addressing fluorogenic real-time qPCR inhibition using the novel custom Excel file system 'FocusField2-6GallupqPCRSet-upTool-001'to attain consistently high fidelity qPCR reactions. Biol Proced Online 8(1):87–153
66. Carcel JA et al (2007) High intensity ultrasound effects on meat brining. Meat Sci 76(4):611–619
67. Gou P, Comaposada J, Arnau J (2003) NaCl content and temperature effects on moisture diffusivity in the Gluteus medius muscle of pork ham. Meat Sci 63(1):29–34
68. Ozuna C et al (2013) Influence of high intensity ultrasound application on mass transport, microstructure and textural properties of pork meat (Longissimus dorsi) brined at different NaCl concentrations. J Food Eng 119(1): 84–93
69. McDonnell C et al (2014) The effect of ultrasonic salting on protein and water–protein interactions in meat. Food Chem 147: 245–251
70. Ojha KS et al (2016) Ultrasound assisted diffusion of sodium salt replacer and effect on physicochemical properties of pork meat. Int J Food Sci Technol 51(1):37–45
71. Inguglia ES et al (2017) Salt reduction strategies in processed meat products—a review. Trends Food Sci Technol 59:70–78

Chapter 9

Pulsed Light for Grape and Wine Processing

Carlos Escott, Iris Loira, and Antonio Morata

Abstract

Pulsed light for sanitation of berries is considered a non-thermal technology able to control the microbial population of ready-to-eat vacuum-packed products. This technology has also been evaluated as pre-treatment for grapes in order to reduce native microorganisms located in the pruina of grapes. The reduction of microorganism counts would allow the use of culture starters to express the floral and fruity character of the yeast-related metabolites produced during winemaking. Here we describe the protocols followed for the treatment of immobilized and non-immobilized destemmed grapes in laboratory-scale PL cabinet and the growth media with the aim of knowing the microbial counts in any time.

Key words Pulsed light, Grape berries, Yeasts, Bacteria, Winemaking, Growth media, Micro-fermentation

1 Introduction

Pulsed light (PL) is a non-thermal emerging technology that uses white light comprising the visible light spectrum, fractions of the ultraviolet, and near-infrared invisible light spectra [1] that can be obtained commonly from inert gases flash lamps filled with xenon or krypton [2]. Regardless of the source, the effects produced by a sequence of high-intensity pulses have been tested in many different industrial fields.

These applications are very diverse and include particles sintering [3], the thermal process of localized surfaces in semiconductors [4], and mainly dermatology applications such as treatments involved in aspects of the skin and photorejuvenation, collagen formation, vascular facial lesions produced from rosacea, facial hemangiomas, stretch marks, and the removal of corporal hair [5–10].

In the food industry, PL has been used as a technology to reduce microbial populations in food stuff since the PL treatment may affect the structure of proteins and the cellular membranes, or it can even promote the destruction of nucleic acid and the

Mohsen Gavahian (ed.), *Emerging Food Processing Technologies*,
Methods and Protocols in Food Science, https://doi.org/10.1007/978-1-0716-2136-3_9,

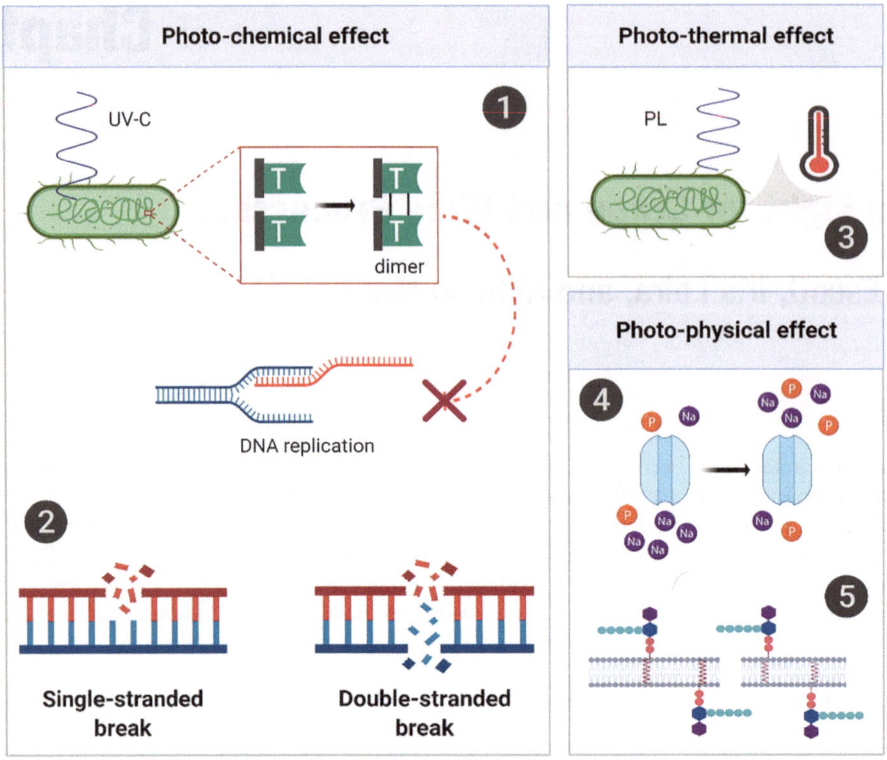

Fig. 1 Effects of UV-C radiation on microorganisms: (1) formation of dimers from adjacent thymines and inhibition of DNA replication, (2) single- and double-stranded breaks of DNA chains, (3) locally increased temperature, (4) ion flow modification, and (5) membrane permeabilization. Taken from [17]

formation of dimers [11]. This property of PL may be an alternative for the reduction of pathogens from food matrices [12, 13].

The energy produced by the ultrashort pulses can be estimated to be several folds the energy of the sun received by the surface of the planet at sea level [14]. The photons released during the emission of the pulses cover a wide range of the electromagnetic spectrum. The UV-C fraction, between 200 and 280 nm, results lethal to microorganisms through the formation of pyrimidine dimers that inhibit the formation of new DNA chains needed for the cell replication [15, 16] and single-strand breaks and double-strand breaks of DNA's structure. There might also be a photo-thermal and photophysical damage in the biological structures of yeast, bacteria, and viruses when using PL. These damages include changes in membrane permeability, depolarization of cell membrane, ion flow variations, and localized heating [3] (Fig. 1).

The control of microorganisms with PL includes, not only, the elimination of food pathogens like *Escherichia coli* [18], *Salmonella enterica* [19], and *Listeria monocytogenes* [20, 21] but also the reduction of spoiling yeasts, molds, and bacteria [22, 23] that may alter fermentative processes as brewing and winemaking.

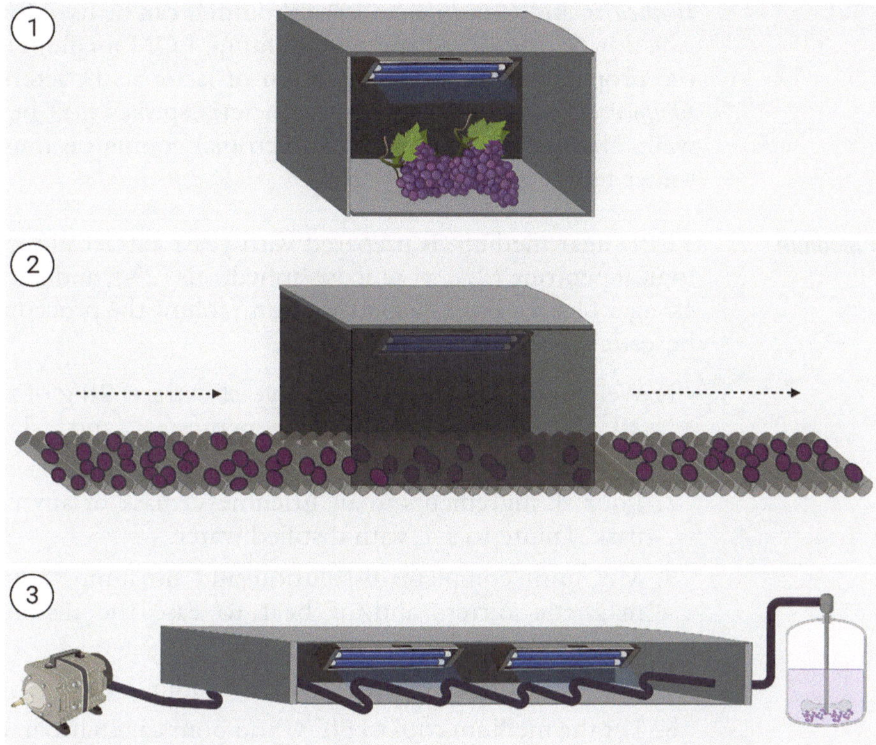

Fig. 2 PL arrangement for different applications: (1) batch-series treatments (laboratory cabinet), (2) proposed continuous treatment on selection table, and (3) proposed continuous treatment for must

For PL technology to be useful in reducing microbial populations in winemaking, the findings obtained at batch-series laboratory scale should be reproducible in a continuous treatment at a selection table and, potentially, also useful in musts after grape crushing (Fig. 2).

This chapter covers the materials and methods followed to assess the use of PL for the reduction of native yeast and bacteria in *Vitis vinifera* varieties and the making of growth solid media and micro-fermentations to evaluate the effectiveness of the technology.

2 Materials

Prepare all media for microbial growth with distilled, double-distilled, or nanopure water to avoid chlorides dissolved. The media described allow the identification of total yeasts and bacteria, or the selective growth of specific genera. YEPD medium allows the growth of all yeast genera. Lysine medium is useful to identify non-*Saccharomyces* yeasts. Chromogenic agar is aimed for isolation and identification of *Candida* species (*Candida albicans*, *Candida tropicalis*, and *Candida krusei*) although it can be used for selective isolation of other yeast species and fungi. LOM medium is used for the propagation and determination of lactic acid bacteria, mainly *Oenococcus oeni*, although other bacteria species may be found as well. The media is suitable for microbial counts on musts, musts under fermentation, and wines.

2.1 YEPD Medium

YEPD agar medium is prepared with yeast extract (1%), bacteriological peptone (2%), D-glucose anhydrous (2%), and bacteriological agar (1.5%). For 1 L solid medium, follow the procedure shown hereafter.

1. Weigh 10 g of bacteriological yeast extract, 20 g of D-glucose, 20 g of bacteriological peptone, and 15 g of bacteriological agar.
2. Pour all ingredients in an Erlenmeyer flask or any volumetric flask. Dilute to 1 L with distilled water.
3. Mix until complete dissolution and no lumps are left. Use magnetic stirrers and/or heat to ease the dissolution (*see* **Note 1**).
4. Autoclave the flask at 121 °C for 15 min.
5. Let the medium cool to 50 °C and pour ca. 25 mL in each Petri plate inside a sterile laminar flow hood or similar.
6. Store Petri plates below 5 °C when possible until further use (*see* **Note 2**).

2.2 Lysine Medium

The lysine medium can be prepared from commercial preparations. This medium contains 1 g/L of L-lysine.

1. Suspend 6.6 g of Lysine medium powder every 100 mL of distilled water.
2. Add 0.84 mL of a potassium lactate solution (60% w/v) per 100 mL.
3. Boil the solution to dissolve the medium completely (*see* **Note 3**).
4. Cool to 50 °C and adjust pH to 5 with a 10% (w/v) solution of L-lactic acid.
5. Pour the medium into Petri plates and store the Petri plates below 5 °C when possible until further use (*see* **Note 2**).

2.3 Chromogenic Medium

The chromogenic agar is a mix of glucose, bacteriological agar, chloramphenicol, chromogenic mixture, and peptone. The procedure described hereafter is for a commercial preparation.

1. Weigh 45.9 g of powder and mix in 1 L distilled or double-distilled water.

2. Continue mixing while heating the solution (*see* **Note 3**).
3. Let the solution boil for at least 1 min until total dissolution of crystals (*see* **Notes 3** and **4**).
4. Let the temperature of the solution decrease to ca. 50 °C.
5. Pour the chromogenic agar into Petri plates inside a sterile laminar flow hood or similar.
6. Store Petri plates below 5 °C when possible until further use (*see* **Note 2**).

2.4 LOM Medium

Commercial LOM medium (*Leuconostoc oenos* medium) contains tryptone, dextrose, yeast extract, fructose, L-cysteine HCL, magnesium sulfate, manganese sulfate, ammonium citrate, and polysorbate. An alternative procedure to prepare LOM medium from zero with fresh sterile tomato serum is shown hereafter.

1. For 1 L solid medium, divide 900 mL into two halves. Weigh the following compounds and dissolve them in the first half:
2. Tryptone 10 g, yeast extract 5 g, glucose 10 g, fructose 5 g, magnesium sulfate ($MgSO_4{\cdot}7H_2O$) 0.2 g, magnesium sulfate ($MnSO_4{\cdot}H_2O$) 0.05 g, diammonium citrate 3.5 g, Tween 80 1 mL, and filtered tomato juice 100 mL.
3. Mix all components until completely dissolved and adjust pH to 4.8.
4. Dissolve 20 g/L of agar powder in the second half of the distilled water.
5. Autoclave both volumes separately at 115 °C for 25 min.
6. Mix and dispense aseptically when still hot.
7. Wait for temperature to cool to 50 °C and add 10 mL of a 5% (w/v) solution of L-cysteine hydrochloride (*see* **Note 5**).
8. Add 2.5 mL of a solution of nystatin 50 g/L (*see* **Note 6**).
9. Mix completely to homogenize the medium.
10. Pour the liquid into Petri plates inside a sterile laminar flow hood or similar.
11. Store Petri plates below 5 °C when possible until further use (*see* **Note 2**).

3 Methods

The methodologies to assess PL in winemaking need the grapes to be destemmed as to emulate the irradiation to be applied in a sterilization area after the selection table (*see* **Note 7**). The protocols are to be carried out at room temperature unless noted otherwise.

Fig. 3 Vacuum-packed grapes in LDPE bags in a single-layer configuration

3.1 Immobilized Grapes

1. Place the grapes inside a plastic bag in a regular one-layer structure (*see* **Note 8**).
2. Use a vacuum apparatus to hermetically seal the bags and fix the grapes in one given position (*see* Fig. 3).
3. Adjust the size of the vacuum-packed grapes to the irradiation zone marked inside the PL chamber (*see* Fig. 4).
4. Place the vacuum-packed grapes on the tray and adjust the height if possible (*see* Fig. 4). The height shall vary in function of the energy density expected. Use an energy dose sensor to know the energy at different heights.
5. Select the power, when possible, or the number of pulses to be applied (*see* **Note 9**).
6. Start the pulses either manually or automatically.
7. Open the chamber and place the vacuum-packed grapes upside down to apply the same treatment on the reverse side (*see* **Note 10**).
8. Place the grapes inside a laminar flow hood and carry on with the *microbial counts* and the *micro-fermentation* methods described further on.

3.2 Non-immobilized Grapes

The purpose of this methodology is to reduce the blind spots created by shadows and also to emulate a continuous PL treatment.

1. Arrange destemmed grapes on a plastic tray in maximum two layers high (*see* Fig. 5) (*see* **Note 11**).
2. Place plastic tray inside PL chamber over the irradiation area to maximize the irradiation (*see* Fig. 4).

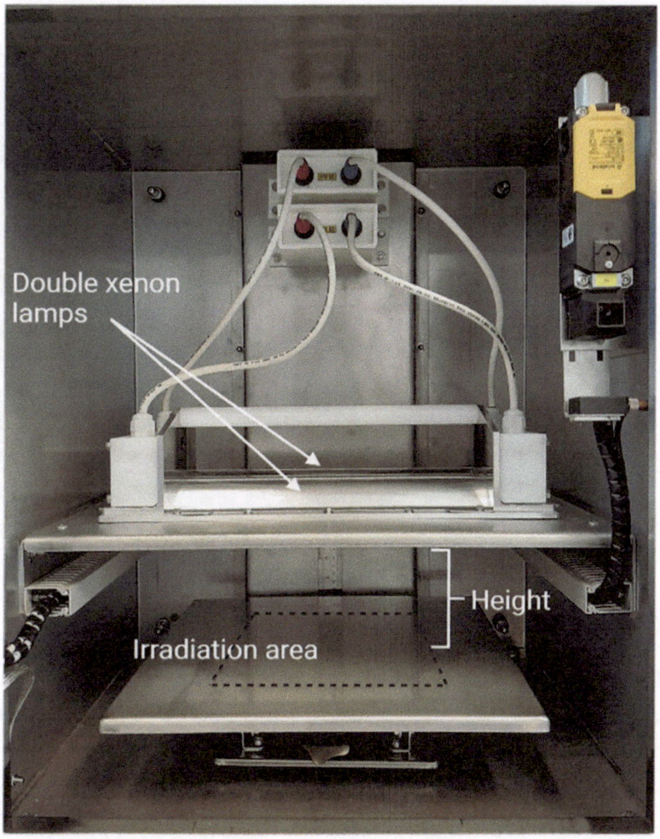

Fig. 4 Inside arrangement of a customized PL chamber. Double xenon lamp, irradiation area, and adjustable height are shown

3. Adjust the height for the grapes to be at 7 cm from the xenon lamps.
4. Select the power, when possible, or the number of pulses to be applied (*see* **Note 9**). For a 2.4 MW PL apparatus, we have selected 5, 10, 20, and 40 pulses (*see* **Note 12**).
5. Remove the tray after the treatment and shake it to randomly change the position of the grapes and simulate the rotation on a conveyor.
6. Place the tray back inside the chamber and repeat the treatment. Repeat this step until three treatments have been applied.
7. Take the tray out and place it inside a laminar flow hood and carry on with the *microbial counts* and the *micro-fermentation* methods described further on.

Fig. 5 Distribution of grapes in one- or two-layer conformation in plastic trays. The size of the trays is 12.5 × 20 cm

3.3 Microbial Counts

1. Crush the grapes after PL treatment in a sterile beaker or similar under sterile environment to avoid cross contamination. Best if done inside a laminar flow hood.
2. Let the must set in contact with the skins and seeds for at least 1 h.
3. Homogenize the must before taking an aliquot of 1 mL with a micro-pipet or 1/5 mL glass pipet.
4. Make tenfold dilutions by adding the 1 mL aliquot into 9 mL sterile nanopure water. Stir perfectly with the use of a vortex mixer.
5. Prepare as many dilutions as needed, especially for inoculated musts where the microbial population can be estimated in 10^7–10^8 CFU/mL.
6. Add 100 μL with a micro-pipet and pour it in every solid growth media and selective media plates. Distribute the liquid with the use of sterile glass beads.
7. Shake plates carefully with the glass beads inside and let dry. Remove glass beads and keep the plates upside down at constant temperature in an electric laboratory heat chamber.
8. Yeasts growth better at 25–28 °C. Bacteria are well kept at 30–35 °C.

3.4 Micro-fermentations

The objective of the micro-fermentations is to evaluate in microscale the fermentative performance of selective yeasts and also to assess the effectiveness of the PL treatment. Small volumes allow to increase the number of trials and to reduce the volume of sample needed.

1. Divide the must in equal volumetric parts in function of the capacity of the fermentation flask. Leave 20% of the volume free as headspace in order to prevent spillage or the block of Müller valves.
2. Prepare the inoculum following the manufacturer indications.
3. Estimate initial population in culture starters following the methodology described previously (*microbial counts*).
4. Inoculate 2% v/v of the culture starter of either single or multiple populations.
5. Register the weight of the fermentation flasks right after the inoculation of the culture starter and continue to record the weight for as long as the fermentation takes place (*see* **Note 13**).
6. Perform microbial counts as deemed appropriate during the fermentation in order to determine yeast implantation.

4 Notes

1. Be careful when heating the YEPD mix with a Bunsen burner due to the sudden increase of volume by the formation of bubbles. The use of heat is recommended when the YEPD medium is to be autoclaved in separate flasks in order to have the medium homogenized before it is divided.
2. It is preferable to use medium casted in Petri dishes within 1 month when kept below 5 °C. The lack of RH can dry the media.
3. It is recommended to use a boiling water bath to ensure that the solution never exceeds 100 °C. The mix can also be heated directly on fire over a Bunsen burner with continuous shacking or stirring. The solution is ready when the lumps disappear and the solution is crystal clear.
4. Be extremely careful when heating over a Bunsen burner to prevent the solution to be overheated or burnt. Avoid autoclaving.
5. L-Cysteine hydrochloride solution (5% w/v) should be sterile by filtration and added to the mix after autoclaving once the temperature has decreased to 50 °C.
6. Prepare the solution of Nystatin 50 g/L in distilled water and sterilize by filtration. Keep the solution in a freezer at −18 °C for better preservation.

7. Using destemmed grapes in the non-immobilized method may induce the release of water from the grapes reducing the effectiveness of the PL treatment. Avoid applying the PL in several cycles as it is better to increase the amount of pulses instead.
8. Use UV transparent plastic bags as much as possible. We recommend the use LDPE bags due to a lower absorption at UV-C range, especially at 254 nm.
9. Commercial PL chambers can have medium and maximum energy modes. Customized and single-mode chambers do not allow to modify the energy reached.
10. The geometry of grapes is prone to producing shadows at the edges. The shadows formed are blind spots for the PL, and thus, the effectiveness of the pulses is less in these areas.
11. There is a marked preference to use plastic trays over metal trays. The tray shall preferably not absorb heat during the treatment to avoid heat transfer between this material and the grapes.
12. The recommended fluence, or amount of energy per surface area treated for effective microbial reduction, shall be between 0.06 and 0.12 kJ/cm^2. This may vary in function of the design of the PL apparatus.
13. The loss in weight recorder in the fermentation flasks is related to the release of CO_2 as the reducing sugars are metabolized by yeasts. The yeast metabolism produces two molecules of ethanol and two molecules of CO_2 out of each hexose consumed. To follow up the weight of the fermentation flasks allows the researcher to estimate the amount of ethanol produced and to know if the fermentation continues or has stopped.

Acknowledgments

This work has been supported by MICINN (Ministerio de Ciencia, Innovación y Universidades - Spain), grant number RTI2018-096626-B-100.

References

1. Barbosa-Canovas GV, Schaffner DW, Pierson MD, Zhang QH (2000) Pulsed light technology. J Food Sci 65:82–85
2. Unni LE, Chauhan OP (2019) Use of pulsed light in food processing. In: Chauhan OP (ed) Non-thermal processing of foods. CRC Press, Boca Raton, pp 173–188. ISBN 9788578110796
3. Ryu J, Kim HS, Hahn HT (2011) Reactive sintering of copper nanoparticles using intense pulsed light for printed electronics. J Electron Mater 40:42–50
4. Kirkpatrick AR (1979) Method involving pulsed light processing of semiconductor devices. Spire Corporation, Bedford, pp 1–8

5. Negishi K, Tezuka Y, Kushikata N, Wakamatsu S (2001) Photorejuvenation for Asian skin by intense pulsed light. Dermatol Surg 27: 627–632
6. Goldberg DJ (2000) New collagen formation after dermal remodeling with an intense pulsed light source. J Cutan Laser Ther 2:59–61
7. Schroeter CA, Haaf-Von Below S, Neumann HAM (2005) Effective treatment of rosacea using intense pulsed light systems. Dermatol Surg 31:1285–1289
8. Angermeier MC (1999) Treatment of facial vascular lesions with intense pulsed light. J Cosmet Laser Ther 1:95–100
9. Hernández-Pérez E, Colombo-Charrier E, Valencia-Ibiett E (2002) Intense pulsed light in the treatment of striae distensae. Dermatol Surg 28:1124–1130
10. Gold MH, Bell MW, Foster TD, Street S (1997) Long-term epilation using the Epi-Light broad band, intense pulsed light hair removal system. Dermatol Surg 23:909–913
11. Pollock AM, Singh AP, Ramaswamy HS, Ngadi M, Singh P (2017) Pulsed light destruction kinetics of L. monocytogenes. LWT Food Sci Technol 84:114–121
12. MacGregor SJ, Rowan NJ, McIlvaney L, Anderson JG, Fouracre RA, Farish O (1998) Light inactivation of food-related pathogenic bacteria using a pulsed power source. Lett Appl Microbiol 27:67–70
13. Rowan NJ, MacGregor SJ, Anderson JG, Fouracre RA, McIlvaney L, Farish O (1999) Pulsed-light inactivation of food-related microorganisms. Appl Environ Microbiol 65:1312–1315
14. Brown AC (2008) Understanding food—principles and preparation, 3rd edn. Thomson Wadsworth, Belmont. ISBN 9780495107453
15. Gómez-López VM, Ragaert P, Debevere J, Devlieghere F (2007) Pulsed light for food decontamination: a review. Trends Food Sci Technol 18:464–473
16. Elmnasser N, Guillou S, Leroi F, Orange N, Bakhrouf A, Federighi M (2007) Pulsed-light system as a novel food decontamination technology: a review. Can J Microbiol 53:813–821
17. Santamera A, Escott C, Loira I, Del Fresno JM, González C, Morata A (2020) Pulsed light: challenges of a non-thermal sanitation technology in the winemaking industry. Beverages 6: 1–16
18. Valdivia-Nájar CG, Martín-Belloso O, Giner-Seguí J, Soliva-Fortuny R (2017) Modeling the Inactivation of *Listeria innocua* and *Escherichia coli* in fresh-cut tomato treated with pulsed light. Food Bioprocess Technol 10:266–274
19. Leng J, Mukhopadhyay S, Sokorai K, Ukuku DO, Fan X, Olanya M, Juneja V (2020) Inactivation of *Salmonella* in cherry tomato stem scars and quality preservation by pulsed light treatment and antimicrobial wash. Food Control 110(107):005
20. Ignat A, Manzocco L, Maifreni M, Bartolomeoli I, Nicoli MC (2014) Surface decontamination of fresh-cut apple by pulsed light: effects on structure, colour and sensory properties. Postharvest Biol Technol 91: 122–127
21. Tao T, Ding C, Han N, Cui Y, Liu X, Zhang C (2019) Evaluation of pulsed light for inactivation of foodborne pathogens on fresh-cut lettuce: effects on quality attributes during storage. Food Packag Shelf Life 21(100):358
22. Agüero MV, Jagus RJ, Martín-Belloso O, Soliva-Fortuny R (2016) Surface decontamination of spinach by intense pulsed light treatments: impact on quality attributes. Postharvest Biol Technol 121:118–125
23. Avalos Llano KR, Marsellés-Fontanet AR, Martín-Belloso O, Soliva-Fortuny R (2016) Impact of pulsed light treatments on antioxidant characteristics and quality attributes of fresh-cut apples. Innov Food Sci Emerg Technol 33: 206–215

Chapter 10

Electrospinning in Food Processing

Conrad O. Perera and Yun Ping Neo

Abstract

The physical phenomenon underlying the electrohydrodynamic atomization also known as electrospinning or electrospraying was first discovered in the seventeenth century by William Gilbert. Although the technology has been applied in the plastic and textile industries, its application in the food industry has been relatively new. Not all biopolymers are electrospinnable; however, cellulose acetate, chitosan, zein, alginate, gelatin, and soy protein have been investigated for the development of edible nanofibers using electrospinning. The factors that govern the electrospinning ability of a polymer are the molecular weight of the polymer, viscosity, conductivity, concentration, and the type of solvent used to dissolve the polymer. Electrospun fibers exhibit high porosity that is deemed suitable as a carrier and an encapsulation method for drugs and therapeutic agents such as antioxidants and other bioactive compounds. This chapter provides a detailed description of the preparation of polymer solutions for electrospinning and control of fiber morphology and diameter, loading of bioactive compounds to polymer solutions, and the resulting fiber properties and stability tests of the loaded nanofibers.

Key words Electrospinning, Electrospraying, Nanofibers, Bioactive compounds, Biopolymers, Encapsulation

1 Introduction

1.1 Historical Background

Electrospinning is a method to produce ultrafine fibers from a polymer solution with diameters ranging from a few nanometers to micrometers using a high-voltage electric field.

In the electrospinning process, a polymer solution is pumped across a high-voltage electric field using a capillary pump. Once the electric field reaches a critical value at which the repulsive electric force overcomes the surface tension of polymer solution, the polymer solution is ejected from the tip to a collector. The polymer jet solidifies due to the fast evaporation of the solvent while traveling to the collector and is deposited on it.

Electrospinning and electrospraying take place under a phenomenon known as electrohydrodynamic atomization (EHDA). The importance of EHDA is that its ability to produce fine, near-

Mohsen Gavahian (ed.), *Emerging Food Processing Technologies*,
Methods and Protocols in Food Science, https://doi.org/10.1007/978-1-0716-2136-3_10,

monodisperse particles and fibers in the micro to nano range, with high precision and efficiency of mass transfer control. The physical phenomenon underlying EHDA was first noted in the seventeenth century by William Gilbert, who observed that a droplet would adopt a cone shape and a spray of liquid would eject from the droplet when a piece of charged amber was placed close to it [1].

In 1882, Lord Rayleigh theoretically estimated the maximum number of charges that a liquid droplet could carry before liquid jets would be ejected from the surface. The physics behind the phenomenon can be summarized as: "the tendency of a capillary force to form a spherical form of equilibrium can be overcome by electrifying the drop" [2]. "Rayleigh instability" was later often denoted as an occurrence of instability of electrospinning fiber jets, which minimizes the surface energy of the liquid jet by breaking it up into droplets [3].

Not all polymers are electrospinnable. The factors that govern the electrospinning ability of a polymer include viscosity, conductivity, molecular weight, and the type of solvent used. The concentration of the polymer used in the solvent determines whether it can be electrospun into nanofibers or not. Concentration also plays a crucial part in the morphology of the fabricated nanofibers [4].

The production of nanofibers from polymer solution/melt under an electrical field has been investigated to produce edible nanofibers for the food industry, biomedical, and cosmetic applications [5]. The low rate of production and the restrictions on the use of solvent that have to comply with food-grade materials are some of the main limitations of electrospinning for the fabrication of edible fibers [6]. Various biopolymers, including cellulose acetate/egg albumen blends [7], chitosan [8], pullulan/β-cyclodextrin blends [9], zein [10], alginate [11], gelatin [12], soy protein [13], etc., have been investigated for the development of edible nanofibers using electrospinning.

1.2 Theory of Electrospinning

During the electrospinning process, the polymer solution is subjected to a strong electric field that causes the buildup of electrical charges within the liquid to form repulsive forces. When the charged electrical forces overcome the surface tension of the polymer liquid, the meniscus of the liquid clinging to the tip of the spinneret will elongate into a conical shape ("Taylor cone") and draws towards the grounded collector [14].

Before reaching the collector, the solvent in the polymer jet evaporates during the flight and the fibers are deposited on the surface of the collector (Fig. 1a). Scanning electron microscopy (SEM) image of electrospun fibers is shown in Fig. 1b. Under certain set of operating conditions, the solution jet breaks down into droplets rather than fibers resulting in electrospraying as shown in Fig. 1c.

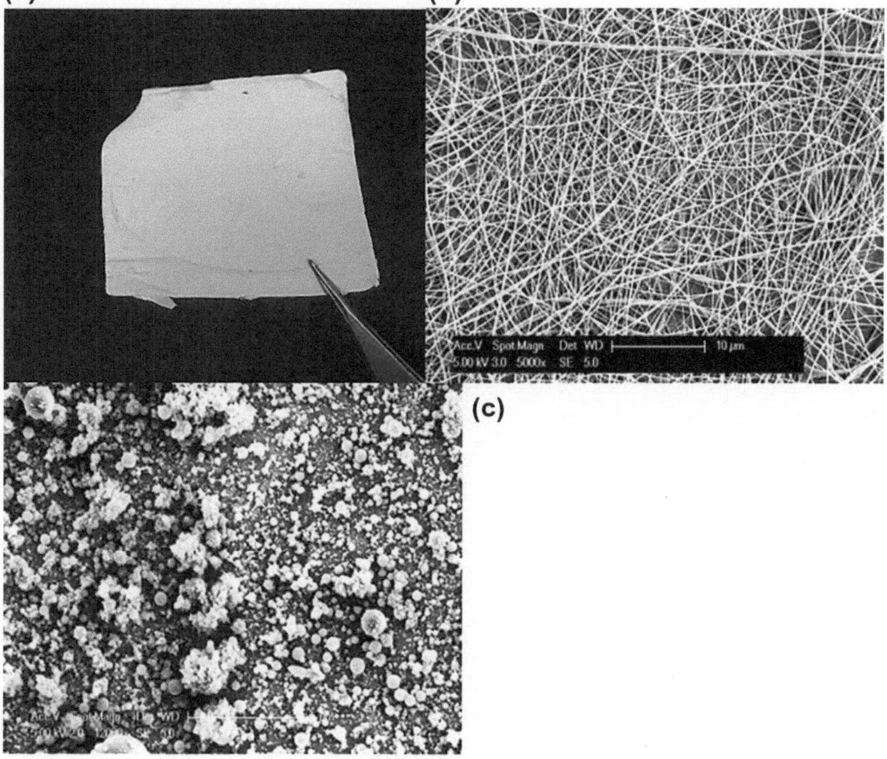

Fig. 1 Image of electrospun fiber mat (**a**); SEM images of electrospun fibers (**b**); electrosprayed droplets (**c**) [15]

2 Materials

2.1 Polymers for Electrospinning

The carriers (polymers) that have been regularly used in the food industry are "generally recognized as safe" (GRAS) materials, which can be natural biopolymers, like polysaccharides, proteins, lipids, and gums, or synthetic polymers, like polyvinyl alcohol (PVOH); polyethylene oxide (PEO); and polyvinylpyrrolidone (PVP) [16]. It is imperative to determine the polymer solution properties such as rheology, viscosity, surface tension, and electrical conductivity to achieve successful electrospinning process. While this information will give a profound insight into the electrospinnability of the selected polymer, it is also important to note that solution properties may vary with the choice of solvent. As a result, each polymer will require certain adjustments together with specific processing conditions in order to be electrospun. Natural biopolymers generally will exhibit limited success in electrospinning due to bio-variations and molecular weight disparity. Consequently, the addition of synthetic polymers into the biopolymers is frequently performed to improve its electrospinnability. Table 1 summarizes

Table 1
Generally recognized as safe (GRAS) polymers for electrospun nanofibers fabrication

Polysaccharides	Protein	Lipid	Synthetic polymers
Cellulose	Collagen	Phospholipid	Polycaprolactone (PCL)
Cellulose acetate	Gelatin		Polyethylene oxide (PEO)
Chitin	Gliadin		Polylactide (PLA)
Chitosan	Hordein		Polyvinyl alcohol (PVOH)
Dextran	Whey, protein isolate		Polyvinylpyrrolidone (PVP)
Ethyl cellulose	Zein		
Guar gum			
Hydroxypropyl methylcellulose			
Hydroxypropyl cellulose			
Methyl cellulose			
Pullulan			
Starch acetate			

various GRAS polymers that have been reported to be electrospinnable in their native forms.

2.2 Electrospinning Setup

A typical electrospinning setup includes a syringe with a metal capillary as the spinneret (e.g., hypodermic syringe needle), syringe pump, high-voltage power supply, and collector that can be grounded or oppositely charged (Fig. 2). The arrangement of the electrospinning setup is generally horizontal, but vertical setups that are controlled by gravitational forces have also been reported. The syringe serves as a reservoir to contain a polymer solution, whereas the pump is used to feed the polymer solution into the spinneret. The spinneret will be charged using a direct current (DC) high-voltage power supply that has either positive or negative outputs. The spinneret can be configured as single, coaxial, or multiaxial to meet different needs. Similarly, the syringe pump can be either single or multichannel that holds more than one syringe. DC high-voltage power supply is the most important component during the electrospinning process. The maximum output of voltage for a DC high-voltage power supply may vary (typically around 20–35 kV). Electrospun fibers will be collected onto a stationary or rotating collector, where the type of collector determines morphology and alignment of the fabricated fibers. The electrospun fiber mats obtained are generally dried in a vacuum oven overnight to remove the residual solvent.

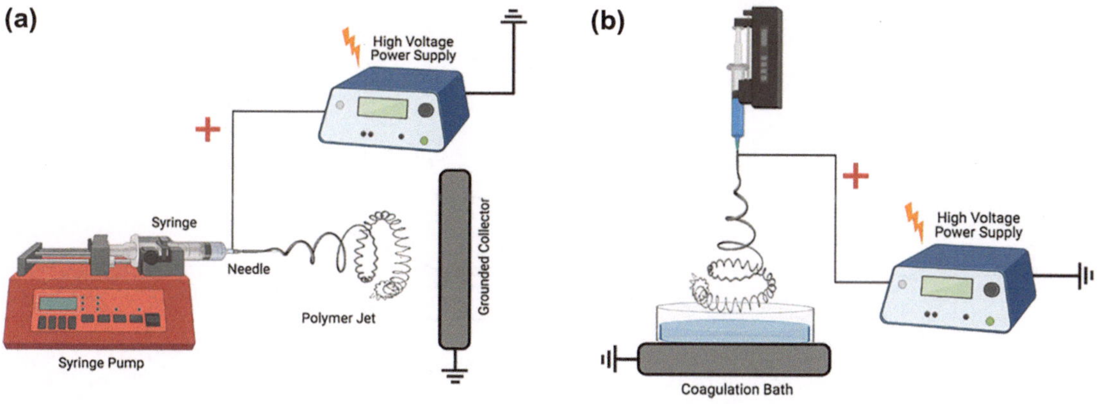

Fig. 2 Schematic illustrations of electrospinning setup. (**a**) Horizontal, (**b**) vertical. (Adapted from Neo [15])

3 Methods

3.1 Preparation of Polymer Solution for Electrospinning

1. When preparing a polymer solution for electrospinning process, it is imperative to make sure that the polymer is fully dissolved in the solvent. In addition, the selected solvent needs to be able to dissolve the targeted bioactive compound for encapsulation without affecting its functional properties. Certain polymers (e.g., chitosan in acetic acid; PVOH in water) might require stirring overnight or heating to yield a uniform solution.
2. The choice of solvent will determine the surface tension, conductivity, and volatility of the solution. Selection of a desirable solvent is generally based on empirical and solubility models. Interestingly, a solvent with high solubility or volatility does not warrant a good electrospinning process. Some studies have proposed the use of ternary solubility diagram (Teas graph) to systematically explore the appropriate electrospinnable binary solvent systems for various polymers [17, 18].
3. Spinning solutions also can be obtained by mixing two polymer solutions prepared separately. Gentle mixing is required to reduce the number of bubbles that may be formed and to obtain a homogeneous solution at a desired ratio.
4. A bare minimum polymer concentration is necessary for fiber formation and varies according to a polymer's molecular weight as well as the type solvent used. The polymer molecular weight will influence the bare minimum polymer concentration for fiber formation. It will also have an effect on viscosity, conductivity, as well as surface tension of the solution.
5. Electrospraying (formation of droplets, a sister technology of electrospinning) will take place when the solution viscosity is too low to sustain a stable electrified polymer jet due to Rayleigh instability. Bead-on-string structure will form at an

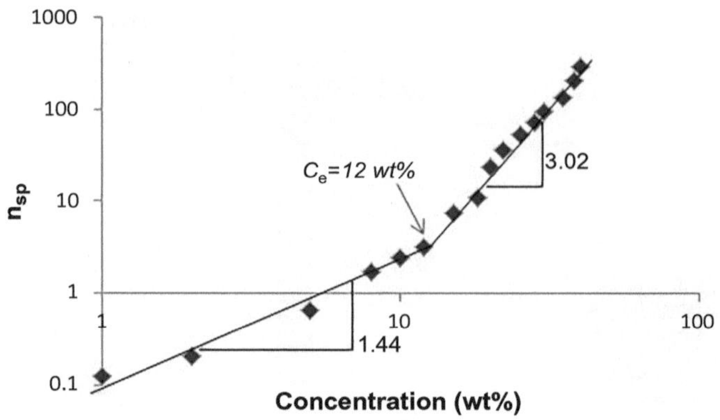

Fig. 3 Logarithmic plot of specific viscosity (η_{sp}) and concentration of zein powder (1–35 wt%) in 80 wt% ethanol aqueous solutions [19]. (Reprinted from Journal of Food Engineering, 109, 645–651, with permission from Elsevier)

intermediate viscosity, which is considered undesirable in electrospinning as it compromises the high surface-area-to-volume ratio. It is useful to know that uniform fibers can be fabricated when the polymer solution is above the entangle concentration (Ce). Hence, the rheological behavior of a polymer solution is deemed important for a successful electrospinning process. Rheological measurements allow the determination of the intrinsic viscosity and molecular organization of the polymer in a solution.

6. A polymer's intrinsic viscosity in a specific solvent can be measured using rotational or microfluidic rheometer at a range of concentrations (C). When the polymer concentration C is less than its entanglement concentration C_e, the polymer solution can be specified as dilute ($C < C_e$), when C is equal to C_e ($C = C_e$), it is considered semi-dilute, and when $C > C_e$, the polymer solution is considered concentrated. C_e can be estimated from a logarithmic plot of specific viscosity against concentration (Fig. 3). A polymer concentration in solution higher than C_e is known to exhibit sufficient chain entanglement for stable electrospinning process to fabricate bead-free fibers [20].
7. Biopolymers with polyelectrolyte character such as alginate and pectin and gums like Arabic gum and Karaya gum are commonly unable to form stable polymer jets during the electrospinning process due to their high surface tension, limited viscoelasticity, and insufficient molecular entanglement [21, 22]. Electrospinning of these biopolymers is frequently performed with another synthetic polymer such as PEO and PVA that serve as carriers to provide necessary entanglement for electrospinning.
8. Salts and surfactants can be added to modulate the solution conductivity and surface tension.

3.2 Control of Electrospun Fiber Morphology and Diameter

1. In addition to solution parameters, the processing and ambient parameters will also influence morphology of electrospun fibers. Processing parameters involve applied voltage, solution feed rate, and distance between spinneret and collector. Ambient parameters are environmental conditions like relative humidity and temperature where the electrospinning takes place. These parameters are subjected to the electrospinning setup and design.
2. The applied voltage will influence the surface charges on the spinneret and the subsequent polymer jet formation. Additionally, the threshold voltage to initiate cone jet formation is also determined by surface tension and viscosity of the polymer solution. A higher voltage can lead to smaller fiber diameter due to higher repulsive forces that promote stretching and whipping of the jet. On the contrary, higher applied voltage can also lead to larger fiber diameter caused by greater mass flow of the polymer solution.
3. Solution feed rate varies with the inner diameter of the spinneret. A simple hypodermic needle with blunted tip and internal diameter below 2 mm is generally used in conventional electrospinning setup. Thinner fibers may be fabricated at lower feed rate as the solvent has more time to be evaporated, which favors the jet elongation.
4. Similarly, a longer distance between the spinneret and the collector will produce thinner fibers as it will allow longer times for the jet to be stretched and elongated. The distance between spinneret and collector is generally suggested to be in the range of 10–20 cm. This is because the electrostatic forces between two can be reduced by the square of the distance between them.
5. Higher environmental temperatures generally promote thinner fibers due to reduction in surface tension and solution viscosity. Furthermore, provision of higher energy to the polymer molecule at higher temperature will also promote electrical conductivity of the solution.
6. The effect of humidity on electrospinning is often overlooked despite relative humidity has been confirmed to influence morphology and structure of the fabricated fibers. A high relative humidity normally will allow the formation of porous, wrinkled, or rough fibers by the polymer jet through moisture condensation or charge dissipation [23]. This wide range of obtained morphologies has offered several possibilities that are beneficial in encapsulation applications.
7. The effect of processing and ambient parameter on electrospinning may vary from case to case. Modeling the relationship and

optimization of these parameters using response surface methodology (RSM) can be considered useful in obtaining the desired fiber morphology and diameter.

3.3 Loading of Bioactive Compounds into Fibers

3.3.1 Blended Electrospinning

Encapsulation of bioactive compounds is often achieved by mixing the active component and polymer in a common solvent as a blend prior to the electrospinning process. Blended electrospinning is a simple method, in which the bioactive compound can be dissolved directly into the polymer solution. The solution jet will solidify and dry up during its flight towards the electrospinning collector that results in entrapment of the active component within the polymeric matrix.

Blended electrospinning is not suitable for the loading of sensitive biological compounds. Examples like proteins, enzymes, or cells may be denatured in the organic solvent used or destroyed during the blending process. Nevertheless, these limitations can be overcome by first encapsulating the sensitive biological compound as small particles followed by blending with polymer solution before electrospinning.

Non-electrospinnable natural polymers such as keratin and lignin could be made into nanofibers by blending them with suitable copolymers, for example, blending lignin with dendrimers such as polyamidoamine (PAMAM). The blending helps lignin to become a more spinnable fiber due to its good miscibility [24].

3.3.2 Coaxial Electrospinning

Coaxial electrospinning involves a concentric spinneret that will form a core-sheath fiber structure. The inner jet or core fluid contains the bioactive compounds, while the outer shell will be an electrospinnable polymer solution.

The bioactive compound and the polymer will be divided in a core-sheath structure through electrospinning of two coaxial capillaries concurrently. This structure governs the release kinetics of the incorporated bioactive component as the release mechanism is led by polymeric shell degradation.

The core and shell solutions are prepared individually in an appropriate solvent followed by loading them into two different syringes. Next, two separate syringe pumps will drive both solutions to a coaxial concentric spinneret that connects both syringes for electrospinning. The electrospinning could be done horizontally or vertically [25].

3.3.3 Emulsion Electrospinning

Emulsion electrospinning is another approach that allows fabrication of core-shell fiber structure. Unlike coaxial electrospinning, this approach involves single feeding capillary that is similar to the blended electrospinning. The bioactive compound and the selected polymer will be dissolved in two immiscible solvents followed by emulsification through the addition of emulsifier or surfactant. By incorporating emulsion droplets, various lipophilic components are allowed to be loaded into the fibers' hydrophilic polymer matrix [26].

For a water-in-oil (w/o) emulsion system, a hydrophilic bioactive compound will be dissolved in aqueous solution as dispersed phase, whereas a hydrophobic polymer together with an emulsifier or surfactant will be dissolved in an organic solvent as continuous phase. The two solutions are homogenized to form a uniform emulsion before the electrospinning process.

The organic continuous phase will evaporate faster than the aqueous dispersed phase during electrospinning process, which resulted in viscosity gradient of the system. The gradient will cause an ongoing development of the aqueous dispersed phase to stay in the fiber interior. Consequently, the hydrophilic bioactive compound can be encapsulated in hydrophobic polymer fiber in core-sheath structure with emulsifier/surfactant at the interface. Due to the high specific surface area, interconnected pore structures, and capability to incorporate active components at nanoscale, electrospun nanofibrous membranes can be used to segregate oil and water at high permeability and low energy cost [27].

3.4 Characterization of Electrospun Fibers

Electrospun fiber mats will exhibit high surface-area-to-volume ratios and quantum confinement effects as compared to its bulk counterparts. Hence, it is essential to perform physicochemical characterization of the fabricated system and to determine the system efficacy. The characterization is usually performed using microscopy (e.g., scanning electron microscopy, transmission electron microscopy), spectroscopy (e.g., X-ray diffraction, infrared spectroscopy), and thermal methods (e.g., differential scanning calorimetry, thermogravimetric analysis). Cross-linking of the electrospun fiber mat will be done to improve their mechanical properties based on the intended application. The bioactivity properties and controlled release performance of the encapsulated bioactive compounds are usually investigated by dissolution of the fibers in a suitable medium followed by spectrophotometric assays.

The electrospun fiber mats exhibit high porosity that is deemed suitable as a drug or therapeutic agent carrier [28]. Electrospinning has been a promising technique for encapsulation of antioxidants and other bioactive compounds [29]. Research on the incorporation of antioxidants and other functional ingredients into electrospun fiber matrices have been reported by several researchers [30–32].

Zein (Ze), a prolamin fraction obtained from maize, is known for its good thermal resistance and oxygen barrier properties, which finds many applications in the packaging sector [33]. Gallic acid is a naturally occurring phenolic acid that is widely present in the plant kingdom. Studies have shown that gallic acid and its derivatives are antioxidants that exhibit anti-inflammatory and antimicrobial abilities [34, 35]. The physicochemical changes of gallic acid-loaded

Table 2
Viscosity and electrical conductivity of Ze and Ze-GA solutions [15]

Gallic acid content (wt% in solid fibers)	Viscosity (mPa s)	Conductivity (mS/cm)
0	165.69 ± 20.4^{b}	1.34 ± 0.08^{b}
5	204.67 ± 44.9^{c}	1.24 ± 0.02^{c}
10	229.10 ± 52.5^{c}	1.17 ± 0.03^{c}
20	263.08 ± 56.8^{d}	1.02 ± 0.07^{d}

Data are displayed in means ± standard deviation of three replications; means in each column bearing different superscripts are significantly different ($p < 0.05$)

zein fibers (Ze-GA) fabricated by using electrospinning as an encapsulation technique have been studied by Neo et al. [30, 31] and Neo [15].

3.4.1 Solution Properties of Zein and Gallic Acid

Zein solutions at a concentration of 25 wt% and gallic acid at different concentrations were prepared by dissolving them in 80 wt% ethanol aqueous solutions.

The viscosity was found to increase with the increment of gallic acid, but the conductivity decreased with the increase in concentration of gallic acid in the polymer solution (Table 2). The conductivity of weak electrolytes such as gallic acid is highly concentration dependent, and the dissociation of weak electrolytes in solution to produce ions is even lower at higher concentrations.

3.4.2 Fiber Properties

The Ze-Ga composite fiber mats were stained with osmium tetroxide and observed using high-resolution TEM (Fig. 4). The neat Ze fibers demonstrated smooth surfaces without any visible aggregates. Additionally, Ze-GA electrospun fibers were found to show dark patches throughout the fibers. A possible explanation for this might be due to staining of the phenol ring of gallic acid with osmium tetroxide that led to formation of black stained areas [36]. This observation is encouraging and implies successful encapsulation of gallic acid within the zein fibers.

3.4.3 Fiber Diameter

There was an increase in average fiber diameter with the increment of incorporated gallic acid content. Such observation is not surprising as the number of molecular entanglement is directly proportional to the gallic acid concentration, which will increase the viscosity and the resultant fiber diameter [19]. Figure 5 shows the distribution of fiber diameters of the Ze-GA composite fibers as the concentration of GA increased from 0% to 20%.

3.4.4 Thermal Properties of the Electrospun Fibers

Figure 6 shows the melting endotherms of the neat zein and Ze-GA composite electrospun fibers. A sharp endothermic peak corresponded to the melting point of highly crystalline gallic acid was

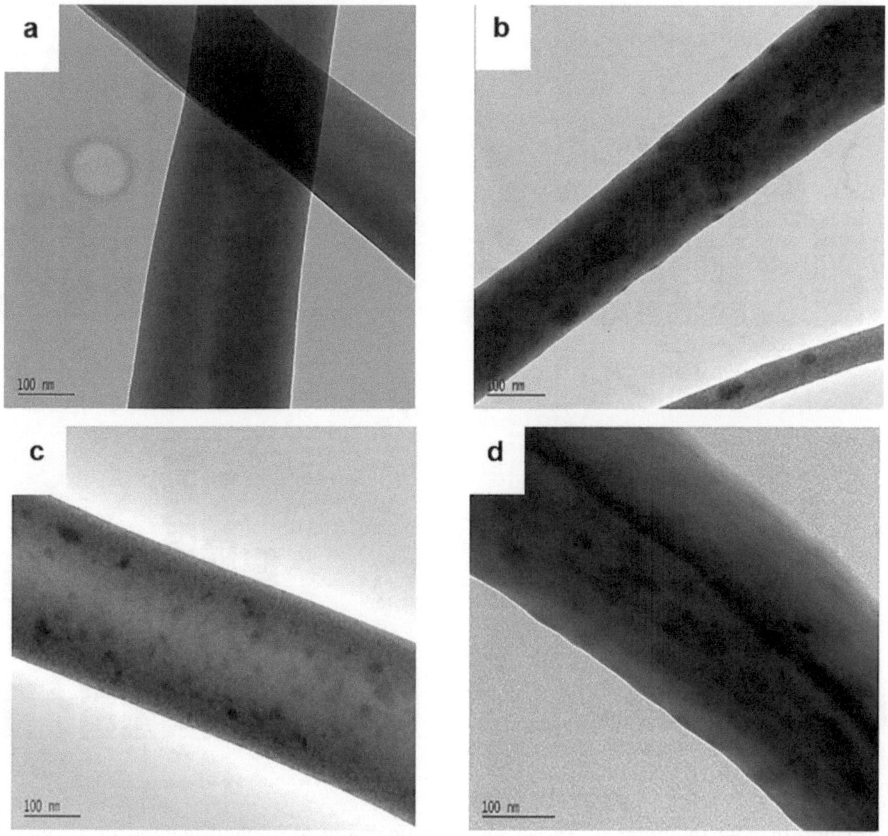

Fig. 4 TEM images of Ze and Ze-GA electrospun fibers at 100 nm scale bar. (**a**) Ze, (**b**) Ze-GA 5%, (**c**) Ze-GA 10%, (**d**) Ze-GA 20% [15]

recorded at 260 °C as shown in Fig. 6e. Interestingly, the neat Ze and Ze-GA composite fibers had demonstrated different thermal profile from pure gallic acid, and a broad endotherm in the range of 40–100 °C with a peak at 68 °C was observed in these electrospun fibers (Fig. 6a–d).

Several studies described these characteristic endotherms of Ze fibers as the "dehydration temperature" (T_D) that is due to the evaporation of bound water or volatile components with elevated temperature [37, 38]. DSC thermograms of the Ze-GA composite fibers (Fig. 6b–d) did not show any melting peaks of gallic acid as seen in Fig. 6e, which suggested a change of physical properties in the loaded gallic acid from highly crystalline to amorphous structure. Glass transition temperature (T_g) of the neat Ze fibers without GA was at 156 °C (Table 3), and degradation of Ze and all the Ze-GA composite fibers took place at approximately 210 °C (Fig. 6). All the Ze-GA composite fibers also exhibited T_g values lower than that of neat Ze in the range of 125–145 °C as shown in Table 3.

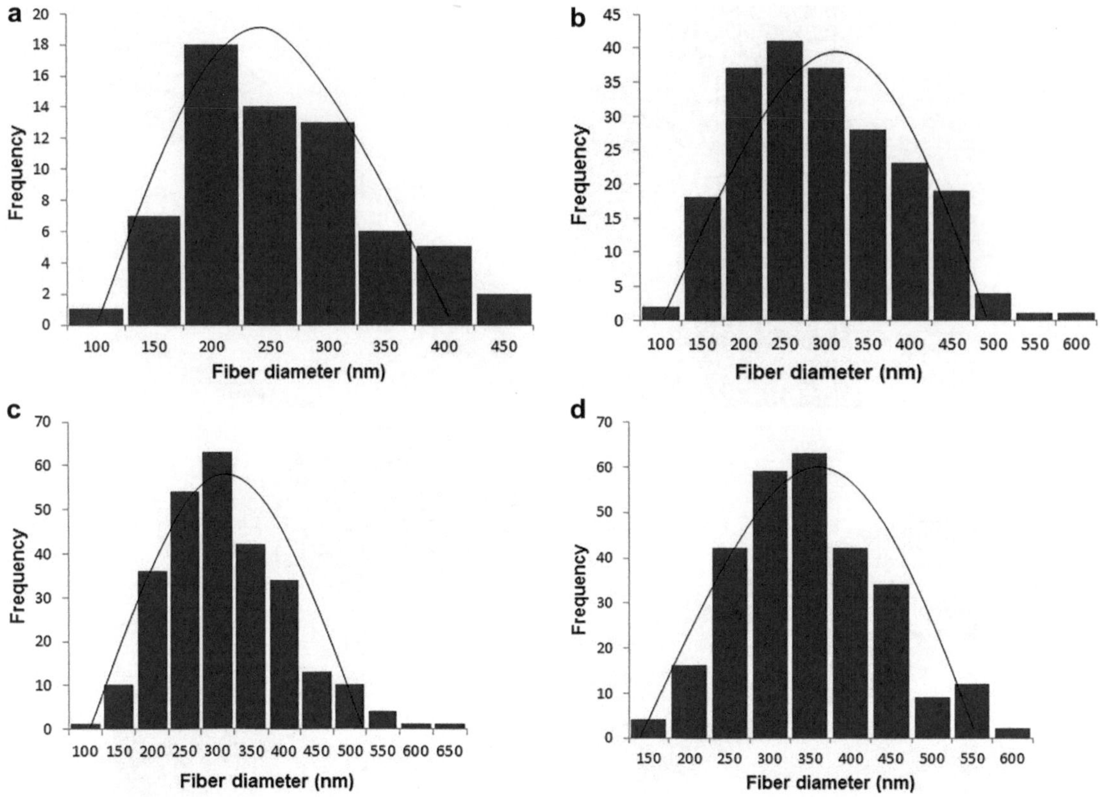

Fig. 5 Distribution of fiber diameter of Ze-GA electrospun fibers at applied voltage of 16 kV, feed rate of 0.8 mL/h, and distance between needle tip and collector of 13 cm. (**a**) Ze, (**b**) Ze-GA 5%, (**c**) Ze-GA 10%, (**d**) Ze-GA 20% [15]

Xie et al. [39] suggested that the reduction of T_g could be due to the plasticizing effect contributed by the incorporated component into the polymer solution [39]. The results obtained also implied the presence of molecular interactions between GA and Ze, where GA has been well integrated within the zein matrices modifying the mobility of zein molecular chains.

3.4.5 X-Ray Diffraction (XRD) Pattern of Electrospun Fibers

Figure 7 is a schematic of X-ray diffraction (XRD) of gallic acid and the electrospun fibers. The XRD studies revealed a high degree of crystallinity of gallic acid, where peaks at 2θ values of 16.1°, 25.3°, and 27.6° were being observed (Fig. 7f). Physical mixture of zein and gallic acid powder (ratio 4:1) was found to exhibit diffraction peaks at around 9.3°, 16.4°, 20.3°, 25.3°, and 27.5° 2θ as shown in Fig. 7b. Peaks at 16.4°, 25.3°, and 27.5° 2θ are characteristics of gallic acid powder in crystalline state. However, a typical amorphous broad peak at around 12.0° 2θ was observed in XRD of the Ze and Ze-GA composite fibers (Fig. 7a, c–e). Interestingly, Ze fibers had lost the two diffraction peaks (Fig. 7a) of raw zein powder as physical mixture at 9.3° and 20.3° 2θ as shown in

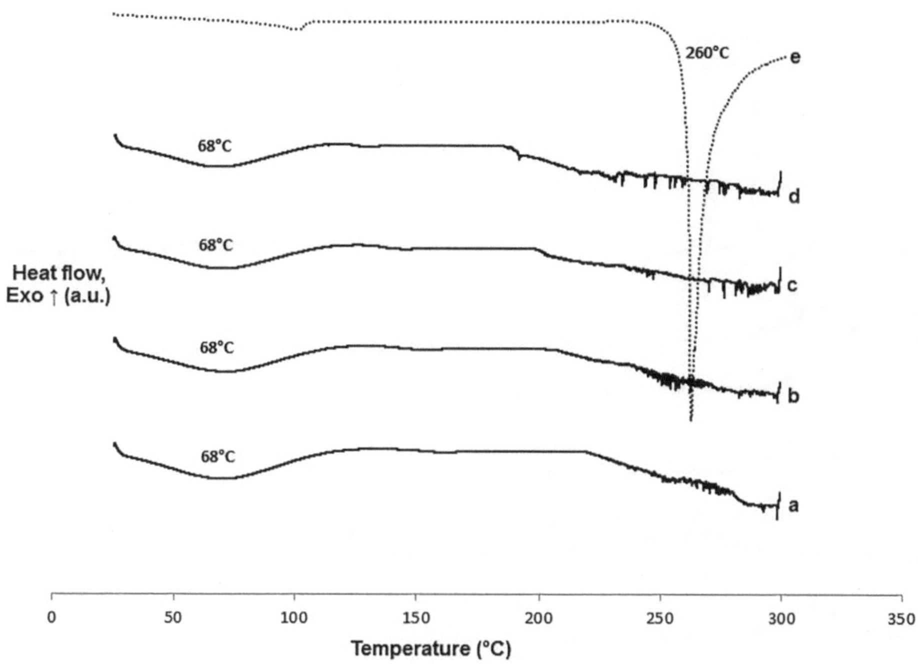

Fig. 6 DSC thermograms of different samples. (a) Ze, (b) Ze-GA 5%, (c) Ze-GA 10%, (d) Ze-GA 20%, (e) gallic acid powder [30]. (Reprinted from Food Chemistry, 136, 1013–1021, with permission from Elsevier)

Table 3
Glass transition temperature (T_g) of Ze-GA electrospun fibers [15]

Gallic acid content (wt% in solid fibers)	T_g (°C)
0	156.30 ± 3.6
5	145.37 ± 3.3
10	137.40 ± 1.9
20	124.99 ± 0.9

Note: Data are displayed in means ± standard deviation of three replications

Fig. 7b, which suggested transition on its physical structure and order after electrospinning. The loss of the signature diffraction peaks of gallic acid powder in the XRD pattern of all Ze-GA electrospun fibers is indicative of difference between the physical structure of gallic acid before and after the incorporation into zein electrospun fibers, which is in line with the findings from DSC studies as highlighted earlier. Shape and position of the amorphous haloes as observed in the Ze-GA composite electrospun fibers were similar to the neat Ze. In general, the composite electrospun fibers had showed amorphous properties similar to neat Ze electrospun fibers even after the incorporation of gallic acid. Diminished crystalline intensity of the gallic acid also may reflect the degree to

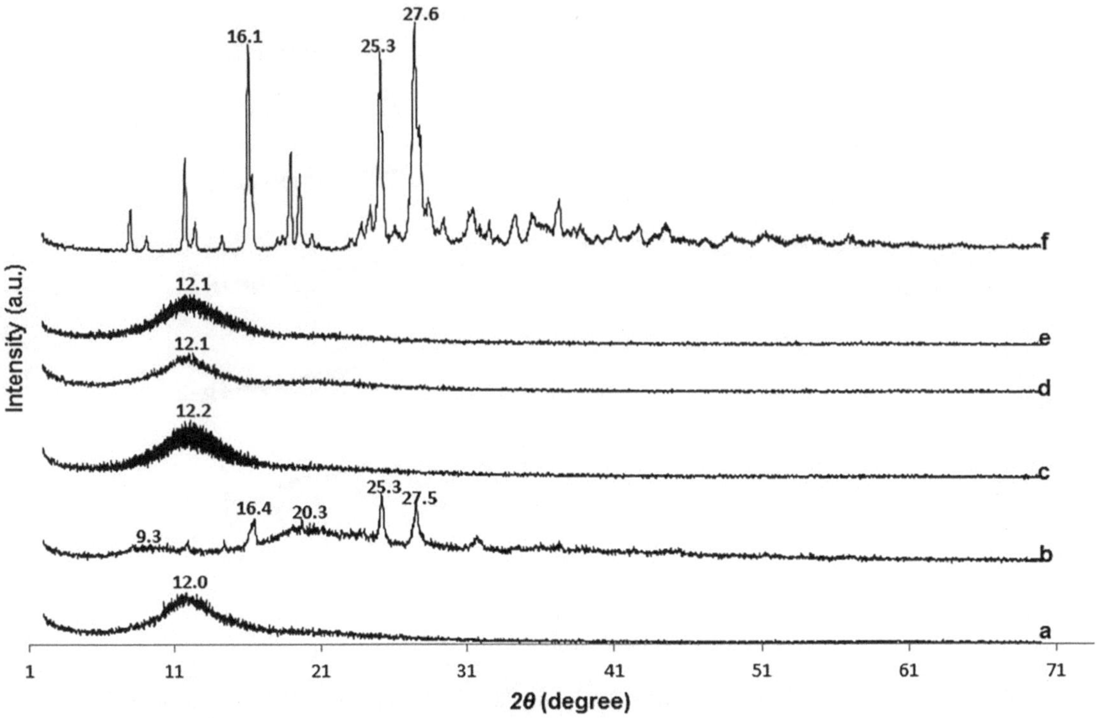

Fig. 7 XRD patterns of different samples. (a) Ze, (b) physical mixture of Ze-GA powder (4:1 w/w), (c) Ze-GA 5%, (d) Ze-GA 10%, (e) Ze-GA 20%, (f) gallic acid powder [30]. (Reprinted from Food Chemistry, 136, 1013–1021, with permission from Elsevier)

which it has been encapsulated within the polymeric network. Rostamabadi et al. [40] suggested that the crystallinity of the fabricated encapsulation system will influence its relative release performance, which signifies the importance of performing X-ray crystallography analysis [40].

3.4.6 Stability Testing Using Attenuated Total Reflectance-Fourier Transform Infrared (ATR-FTIR) Spectroscopy

ATR-FTIR spectroscopy is a good method to measure stability of the nanofiber matrices at high temperatures and storage. In the case of zein and gallic acid nanofiber mats that were described earlier, ATR-FTIR spectra were recorded on a Thermo Electron NICOLET 8700 FTIR spectrophotometer using the Smart Orbit ATR accessory with diamond crystal, single bounce at 45° (Thermo Electron Corporation, Waltham, MA, USA) over the wave number region of 600–3600 cm^{-1}. Interferograms were averaged for 32 scans at 4 cm^{-1} resolution. The measurements were carried out in triplicate, and the signals were processed using the OMNIC 5.0 spectroscopic software (Nicolet Instrument Corp., Madison, WI). The stability test was repeated after storing a new batch of Ze and Ze-GA fiber mats individually at 60 °C for 30 days. The ATR-FTIR spectra were recorded, and the measurements were carried out in duplicate and are shown in Fig. 8. There were no

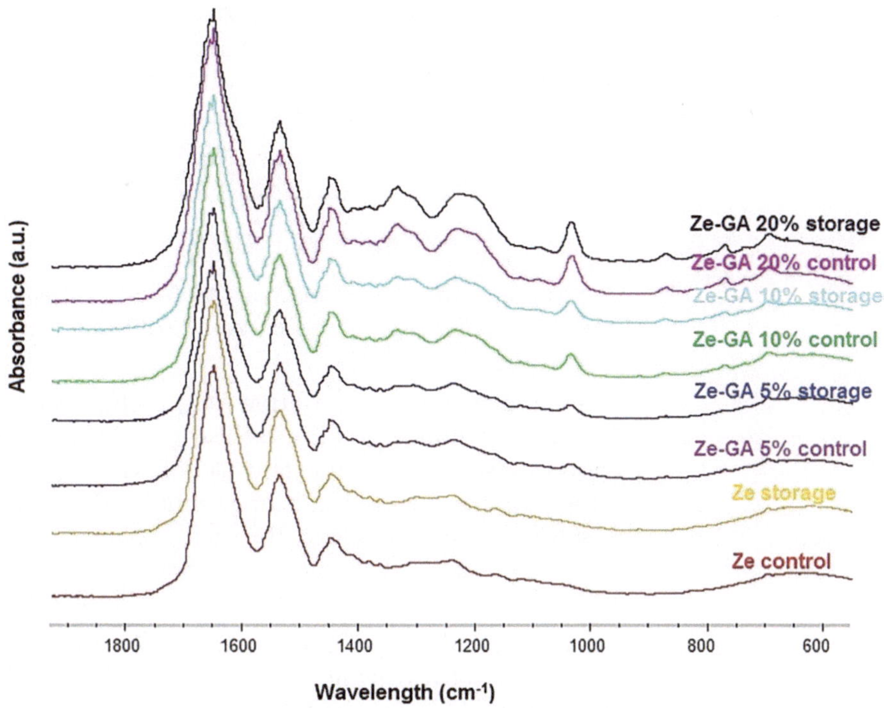

Fig. 8 ATR-FTIR spectra of Ze and Ze-GA electrospun fibers after 30 days of storage at 60 °C [15]

changes in the spectra after storage at 60 °C for 30 days, showing good stability of the zein-gallic acid nanofibers.

4 Notes

1. In general, prepare a range of biopolymer concentrations to be investigated (e.g., 0.1–30%, w/v). Take note on the moisture content of the biopolymer powder when performing the calculations. For each concentration, weigh the biopolymer powder into a right vessel. Add the solvent with constant stirring on a magnetic stirrer until the biopolymer powder is fully dissolved. Depending on the polymer and its solubility in the solvent, the samples may need to be heated for about 1 h or more to facilitate the process. Turn off heat and allow the solution to cool to room temperature. The solution is then ready for characterization such as rheological testing and electrospinning.
2. In preliminary experiments, it was found that Ze solutions of over 25 wt% concentration gave uniform nanofibers without blemishes [15]. Zein solutions used for electrospinning were prepared at a concentration of 25 wt% by dissolving Ze powder in 80 wt% ethanol aqueous solutions under constant stirring at

room temperature. Electrospun Ze fibers containing various amounts of gallic acid were prepared by first dissolving gallic acid in 80 wt% ethanol aqueous solutions, followed by dissolving Ze powder to obtain 25 wt% Ze solutions.

3. The specific viscosity η_{sp} of the polymer was calculated from Eq. 1.

$$\eta_{sp} = (\eta_0 - \eta_s)/\eta_s \quad (1)$$

where η_s is the solvent viscosity (2.58 mPa s), η_0 is the zero shear rate viscosity of the zein solutions at different concentrations.

4. The entanglement concentration (C_e) is the concentration at which a logarithmic plot of zero shear viscosity against concentration intersects. It is important to note that C_e is dependent on the polymer chain length, macromolecular structure, and the solvent used. Therefore, for each polymer in a given solvent, the C_e has to be determined first in order to produce uniform nanofibers with the desired morphological properties.

5. Instead of DC high-voltage power supply, AC potential can also be used as the driving force in electrospinning. A study by Kessick et al. [41] suggested that the electrospun fibrous matrices fabricated using an AC potential would exhibit a greater degree of fiber alignment and more residual solvent as compared to DC potential [41].

6. The major drawback of blended electrospinning is migration of the charged bioactive molecules onto the jet surface during electrospinning process. The bioactive compounds are randomly dispersed in the fiber matrix and generally observed on the surface of the fibers. Localization of the bioactive compounds on the fiber surface will lead to unwanted burst of release. Burst release may be resolved by inducing chemical bonding between the bioactive compounds with the polymeric carrier at molecular level. Additionally, the release performance can also be improved by blending cyclodextrin inclusion complexes of the bioactive compound with the polymer solution before electrospinning process [42].

7. Coaxial electrospinning is a complex technique that requires careful control of processing parameters. A common solvent is suggested to be added into the core and shell fluids to attain the best electrospinning results [43]. The flow rate of both inner and outer fluids has to be precisely tailored in order to achieve stable coaxial jet formation. Successful fabrication of core-sheath structure using an inner needle diameter of 0.21–0.35 mm and an outer needle of 0.82–1.2 mm with core-to-shell flow ratios of 1:3 and 1:10 was reported by William et al. [44].

8. Viscosity of the dispersed and continuous phase; electrospinnability of the polymer; type of emulsifier/surfactant; and processing conditions need to be carefully tuned for successful emulsion electrospinning. Viscosity of the dispersed phase is required to be higher than continuous phase in order to form a continuous core along the fiber axis [45] as suggested by Angeles et al. [45]. Emulsion electrospinning is more likely to destruct the sensitive bioactive compounds as compared to coaxial electrospinning due to the mixing and interface tension between aqueous and organic phase of the emulsion [46].

References

1. Luo CJ, Loh S, Stride E, Edirisinghe M (2012) Electrospraying and electrospinning of chocolate suspensions. Food Bioprocess Technol 5: 2285–2300
2. Rayleigh L (1882) XX. On the equilibrium of liquid conducting masses charged with electricity. Philos Mag 14:184–186
3. Miao J, Miyauchi M, Simmons TJ, Dordick JS, Linhardt RJ (2010) Electrospinning of nanomaterials and applications in electronic components and devices. J Nanosci Nanotechnol 10: 5507–5519
4. Muthu Kumar TS, Senthil Kumar K, Rajini N, Siengchin S, Ayrilmis N, Rajulu AV (2019) A comprehensive review of electrospun nanofibers: food and packaging perspective. Compos Part B 175(2019):107074. https://doi.org/10.1016/j.compositesb.2019.107074
5. Kouhi M, Prabhakaran MP, Ramakrishna S (2020) Edible polymers: an insight into its application in food, biomedicine and cosmetics. Trends Food Sci Technol 103(2020): 248–263
6. Kouhi M, Jayarama Reddy V, Ramakrishna S (2019) GPTMS-modified bredigite/PHBV nanofibrous bone scaffolds with enhanced mechanical and biological properties. Appl Biochem Biotechnol 188:357–368
7. Wongsasulak S, Patapeejumruswong M, Weiss J, Supaphol P, Yoovidhya T (2010) Electrospinning of food-grade nanofibers from cellulose acetate and egg albumen blends. J Food Eng 98:370–376
8. Yue T-T, Li X, Wang X-X, Yan X, Yu M, Ma J-W et al (2018) Electrospinning of carboxymethyl chitosan/polyoxyethylene oxide nanofibers for fruit fresh keeping. Nanoscale Res Lett 13:239. https://doi.org/10.1186/s11671-018-2642-y
9. Fuenmayora CA, Mascheronia E, Cosioa MS, Piergiovannia L, Benedettia S, Ortenzic M, Schiraldia A, Manninoa S (2013) Encapsulation of R-(+)-limonene in edible electrospun nanofibers. Chem Eng Trans 32:1771–1776. https://doi.org/10.3303/CET1332296
10. de Oliveira Mori CLS, dos Passos NA, Oliveira JE, Mattoso LHC, Mori FA, Carvalho AG et al (2014) Electrospinning of zein/tannin bio-nanofibers. Ind Crop Prod 52:298–304
11. Rezaei S, Valipouri A, Ravandi SAH, Kouhi M, Mobarakeh LG (2019) Fabrication, characterization, and drug release study of vitamin C–loaded alginate/polyethylene oxide nanofibers for the treatment of a skin disorder. Polym Adv Technol 30:2447–2457
12. Tang Y, Zhou Y, Lan X, Huang D, Luo T, Ji J et al (2019) Electrospun gelatin nanofibers encapsulated with peppermint and chamomile essential oils as potential edible packaging. J Agric Food Chem 67(8):2227–2234
13. Shankar A, Seyam A-F, Hudson S (2013) Electrospinning of soy protein fibers and their compatibility with synthetic polymers. J Textile Apparel Technol Manag 8(1):1–14. Spring 2013
14. Hohman MH, Shin M, Brenner MP (2001) Electrospinning and electrically forced jets. I. Stability theory. Phys Fluids 13(8): 2201–2220
15. Neo YP (2014) Electrospinning as a novel encapsulation method for food applications. PhD thesis, The University of Auckland, New Zealand
16. Zagho MM, Elzatahry A (2016) Recent trends in electrospinning of polymer nanofibers and their applications as templates for metal oxide nanofiber preparations. In: Electrospinning - material, techniques, and biomedical applications. IntechOpen, London. https://doi.org/10.5772/65900
17. Lasprilla-Botero J, Álvarez-Láinez M, Lagaron JM (2018) The influence of electrospinning

parameters and solvent selection on the morphology and diameter of polyimide nanofibers. Mater Today Commun 14:1–9

18. Luo CJ, Nangrejo M, Edirisinghe M (2010) A novel method of selecting solvents for polymer electrospinning. Polymer 51(7):1654–1662
19. Neo YP, Ray S, Easteal AJ, Nikolaidis MG, Quek SY (2012) Influence of solution and processing parameters towards the fabrication of electrospun zein fibers with sub-micron diameter. J Food Eng 109:645–651
20. Kong L, Ziegle GR (2014) Molecular entanglement and electrospinnability of biopolymers. J Vis Exp (91):e51933. https://doi.org/10.3791/51933. http://www.jove.com/video/51933
21. Padil VV, Cheong JY, AkshayKumar KP, Makvandi P, Zare EN, Torres-Mendieta R, Wacławek S, Černík M, Kim ID, Varma RS (2020) Electrospun fibers based on carbohydrate gum polymers and their multifaceted applications. Carbohydr Polym 247(1):116705
22. Xiao Y, Lin L, Shen M, Shi X (2020) Design of DNA aptamer-functionalized magnetic short nanofibers for efficient capture and release of circulating tumor cells. Bioconjug Chem 31(1):130–138
23. Szewczyk PK, Stachewicz U (2020) The impact of relative humidity on electrospun polymer fibers: from structural changes to fiber morphology. Adv Colloid Interf Sci 286:102315. https://doi.org/10.1016/j.cis.2020.102315
24. Akbari S, Bahi A, Farahani A, Milani AS, Ko F (2021) Fabrication of lignin/dendrimer electrospun blended fiber mats. Molecules 26(3):518. https://doi.org/10.3390/molecules26030518
25. Baykara T, Taylan G (2021) Coaxial electrospinning of PVA/Nigella seed oil nanofibers: processing and morphological characterization. Mater Sci Eng B 265:115012. https://doi.org/10.1016/j.mseb.2020.115012
26. Arecchi A, Mannino S, Weiss J (2010) Electrospinning of poly(vinyl alcohol) nanofibers loaded with hexadecane nanodroplets. J Food Sci 75:N80–N88
27. Su R, Li S, Wu W, Song C, Liu G, Yu Y (2021) Recent progress in electrospun nanofibrous membranes for oil/water separation. Sep Purif Technol 256(2021):117790. https://doi.org/10.1016/j.seppur.2020.117790
28. Deng Y-F, Zhang D, Zhang N, Huang T, Lei Y-Z, Wang Y (2021) Electrospun stereocomplex polylactide porous fibers toward highly efficient oil/water separation. J Hazard Mater 407:124787. https://doi.org/10.1016/j.jhazmat.2020.124787
29. Celebioglu A, Uyar T (2021) Electrohydrodynamic encapsulation of eugenol-cyclodextrin complexes in pullulan nanofibers. Food Hydrocoll 111:106264. https://doi.org/10.1016/j.foodhyd.2020.106264
30. Neo YP, Ray S, Jin J, Gizdavic-Nikolaidis M, Nieuwoudt MK, Liu D, Quek SY (2013a) Encapsulation of food grade antioxidant in natural biopolymer by electrospinning technique: a physicochemical study based on zein–gallic acid system. Food Chem 136:1013–1021
31. Neo YP, Swift S, Ray S, Gizdavic-Nikolaidis M, Jin J, Perera CO (2013b) Evaluation of gallic acid loaded zein sub-micron electrospun fiber mats as novel active packaging materials. Food Chem 141:3192–3200
32. Hu X, Wang X, Han L, Li S, Zhou W (2021) Antioxidant and antimicrobial polyvinyl alcohol electrospun nanofibers containing baicalein-hydroxypropyl-β-cyclodextrin inclusion complex. Colloids Surf A Physicochem Eng Asp 614:126135. https://doi.org/10.1016/j.colsurfa.2021.126135
33. Zhou L, Wang Y (2021) Physical and antimicrobial properties of zein and methylcellulose composite films with plasticizers of oleic acid and polyethylene glycol. LWT Food Sci Technol 140:110811. https://doi.org/10.1016/j.lwt.2020.110811
34. Ahmad B, Hafeez N, Rauf A, Bashir S, Linfang H, Rehman M, Mubarak MS, Uddin MS, Bawazeer S, Shariati MA, Daglia M, Wan C, Rengasamy KRR (2021) *Phyllanthus emblica*: a comprehensive review of its therapeutic benefits. S Afr J Bot 138(202):278–310. https://doi.org/10.1016/j.sajb.2020.12.028
35. Rodrigues L, Donado-Pestana CM, Moura MHC, Rossi e Silva R, Pessoa EVM, Genovese MI (2021) Phenolic compounds from jaboticaba (*Plinia jaboticaba* (Vell.) Berg) ameliorate intestinal inflammation and associated endotoxemia in obesity. Food Res Int 141:110139. https://doi.org/10.1016/j.foodres.2021.110139
36. Kriegel C, Kit KM, McClements DJ, Weiss J (2009) Nanofibers as carrier systems for antimicrobial microemulsions. Part I: Fabrication and characterization. Langmuir 25:1154–1161
37. Luo Y, Teng Z, Wang Q (2011) Development of zein nanoparticles coated with carboxymethyl chitosan for encapsulation and controlled release of vitamin D3. J Agric Food Chem 60:836–843
38. Muller V, Piai JF, Fajardo AR, Favaro SL, Rubira AF, Muniz EC (2011) Preparation and

characterization of zein and zein-chitosan microspheres with great prospective of application in controlled drug release. J Nanomater 2011:928728. https://doi.org/10.1155/2011/928728

39. Xie C, Li X, Luo X, Yang Y, Cui W, Zou J, Zhou S (2010) Release modulation and cytotoxicity of hydroxycamptothecin-loaded electrospun fibers with 2-hydroxypropyl-β-cyclodextrin inoculations. Int J Pharm 391:55–64
40. Rostamabadi H, Falsafi SR, Assadpour E, Jafari SM (2020) Evaluating the structural properties of bioactive-loaded nanocarriers with modern analytical tools. Compr Rev Food Sci Food Saf 19(6):3266–3322
41. Kessick R, Fenn J, Tepper G (2004) The use of AC potentials in electrospraying and electrospinning processes. Polymer 45(9):2981–2984
42. Oliveira MF, Suarez D, Rocha JCB, Teixeira AVND, Cortes ME, De Sousa FB, Sinisterra RD (2015) Electrospun nanofibers of polyCD/PMAA polymers and their potential application as drug delivery system. Mater Sci Eng C Mater 54:252–261
43. Nikmaram N, Roohinejad S, Hashemi S, Koubaa M, Barba FJ, Abbaspourrad A, Greiner R (2017) Emulsion-based systems for fabrication of electrospun nanofibers: food, pharmaceutical and biomedical applications. RSC Adv 7(46):28951–28964
44. Williams GR, Raimi-Abraham BT, Luo CJ (2018) Nanofibers in drug delivery. UCL Press, London. https://doi.org/10.14324/111.9781787350182. 212 p
45. Angeles M, Cheng HL, Velankar SS (2008) Emulsion electrospinning: composite fibers from drop breakup during electrospinning. Polym Adv Technol 19(7):728–733
46. Zamani M, Prabhakaran MP, Ramakrishna S (2013) Advances in drug delivery via electrospun and electrosprayed nanomaterials. Int J Nanomedicine 8:2997–3017. https://doi.org/10.2147/IJN.S43575

Chapter 11

Methods for Screening and Evaluation of Edible Coatings with Essential Oils as an Emerging Fruit Preservation Technique

Kataryne Árabe Rimá de Oliveira, Louise Iara Gomes de Oliveira, Evandro Leite de Souza, and Marciane Magnani

Abstract

Use of essential oils (EO) in the formulation of edible coatings has been considered an emerging technology for controlling fruit postharvest decay. Among the materials used to produce edible coatings, chitosan (Chi) is the most commonly used for EO incorporation. This book chapter describes the procedures used for the preparation of Chi and EO dispersions, as well as for in vitro assays to evaluate the effects of these substances against fungi associated with fruit postharvest diseases. Described methods include the determination of effects on fungal spore germination, fungal mycelial growth, and type of interaction between Chi and EO applied in combination. In addition, methods to evaluate the efficacy of formulated coatings when applied on fruit to prevent (preventive effect) and control (curative effect) fungal postharvest diseases, as well as to measure disease incidence (% of fruit affected of a specific disease) and severity (lesion diameter), are detailed. Tips to improve the repeatability of results are shown, and protocols are not limited to fruit, which is mentioned in this chapter.

Key words Fruit decay, Phytopathogen, Chitosan dispersion, Essential oil emulsion, Coating, Antifungal effects, Preventive assay, Curative assay

1 Introduction

Fruits are submitted to an extensive postharvest handling chain often resulting in quality losses and short shelf life mainly related to occurrence of physiological disorders, mechanical injuries, and action of phytopathogens that cause infection and decay [1].

Low temperatures, controlled/modified atmosphere packaging, irradiation, and chemical fungicides have been used to extend the shelf life and control postharvest fungal infections in fruit [2]. However, the application of some of these strategies may injury quality characteristics of the fruit, not being considered for practical use. In addition, excessive and continuous use of chemical

Mohsen Gavahian (ed.), *Emerging Food Processing Technologies*,
Methods and Protocols in Food Science, https://doi.org/10.1007/978-1-0716-2136-3_11,

fungicides has been a concern because of their association with development of resistance in pathogenic fungi and potential harmful effects to environment and human health [3, 4].

Use of edible coatings with natural antimicrobial substances as alternative technology to reduce the occurrence of postharvest fungal diseases in fruit has emerged as an environmentally friendly option [3].

Edible coatings are characterized as a thin and translucent layer of biodegradable materials (polysaccharide, protein, and lipid) self-assembled on the surface of fruit after application in liquid form by immersing, spraying, dripping, or brushing [5, 6]. Application of edible coatings in fruit has been associated with technological advantages such as antimicrobial and antioxidants effects, reduction in water loss, respiration rate, and inhibition of ethylene production [7].

Chitosan (Chi) and essential oils (EO) are generally recognized as safe (GRAS) by the FDA, and their combined use may enhance their efficacy to preserve fruit [8]. Chi (2-amino-2-deoxy-D-glycopyranose) is a cationic biopolymer derived from deacetylation of chitin in alkaline media (2-acetamide-2-deoxy-D-glycopyranose), with low toxicity to human and film forming ability in acid medium [9]. Chi coatings are biodegradable and semipermeable to gases (CO_2 and O_2) and have inhibitory effects on a variety of postharvest fungi [3, 4, 8, 10].

Antifungal properties of Chi have been related to electrostatic interaction between its cationic chain (amino groups) and negatively charged microbial cell membranes, causing alterations in membrane permeability and loss of electrolytes and other intracellular constituents needed for fungal growth and survival [11]. It has been also suggested that Chi could bound with DNA, causing mRNA synthesis inhibition, enzymatic activity disturbance, and decreased fungal growth [11].

Incorporation of EO in Chi coatings improves significantly the antifungal and antioxidant effects and permits reductions in the doses required to inhibit target organisms in fruit [4, 6]. However, the evaluation of possible interactions (antagonistic, additive, and/or synergistic) becomes the primarily step in an approach to apply Chi and selected EO in combination [4].

EO are volatile and bioactive materials naturally produced as secondary metabolites in different organs of aromatic plants [4]. These compounds have low resistance-inducing effects because they are characterized by a mixture of distinct constituents with varied action modes on target microorganisms [7].

Antifungal properties of EO could be associated primarily with their abilities to cause disorganization of fungal cells membranes structures, inducing depolarization and physical or chemical alterations and disrupting various metabolic activities and enzymatic reactions necessary for mycelial growth/hyphal morphogenesis [4, 5].

One of the main limitations for practical application of EO should be their typical strong odor and taste that may affect negatively the fruit sensory characteristics [8]. Consequently, combined use of low effective antifungal doses of Chi and EO could be employed to decrease negative sensory impacts on coated fruit [4]. In addition, when incorporated in Chi coatings, EO may be continuously released on fruit surface during storage [12].

This chapter provides detailed protocols for preparation of Chi and EO dispersions/coatings (individually or in combinations) and in vitro assays to evaluate the inhibitory effects of these substances against different fungi associated with fruit postharvest losses. Additionally, methods to evaluate the efficacy of formulated coatings when applied on fruit to prevent (preventive effect) and control (curative effect) the development of fungal postharvest diseases are detailed.

2 Materials

2.1 Fungal Strains and Growth Conditions

Select fungal strain related to fruit diseases is an important experimental criterion (*see* **Note 1**).

1. Culturing tools: pipettes, laboratory glassware, plastic 1.5 mL micro tubes, and Petri dishes (90 mm diameter).
2. Prepare selected culture medium according to manufacturer instructions provided on the label. As a general rule, it is wise to prepare just enough medium for each test. Weigh the amount of powder (dehydrated medium) accurately using an analytical balance and hydrate the medium with distilled/deionized water. Make the complete dissolution with the aid of a glass stem or brief warming. Sterilize with autoclaving at 121 °C, 1 atm for 15 min. Distribute approximately 15 mL of cooled sterilized medium in Petri dish (*see* **Note 2**).
3. 1% (w/v) sterile saline solution. Weigh 1 g of sodium chloride (NaCl) for each 100 mL of solution. Transfer to a glass container. Add distilled/deionized water. Shake with the aid of a glass stem up to complete dissolution and autoclave (120 °C, 1 atm, 15 min). Properly identify, date, and store.
4. BOD incubator (biological oxygen demand) of high degree temperature accuracy.

2.2 Preparation of Chi and EO Dispersions and Coatings

1. Laboratory tools: automatic micropipettes, pipettes, and laboratory glassware.
2. Chi dispersions: Chi of medium molecular weight (extracted from shrimp cells or produced by biotechnological processes, deacetylation degree ≥75%) (*see* **Notes 3** and **4**).

3. 3 M sodium hydroxide (NaOH) solution. Check the purity of NaOH (% w/w) on the packaging label. Calculate the mass of NaOH needed to prepare 100 mL of 3 M (mol/L) solution. On an analytical balance, weigh the NaOH and add to a glass container with 100 mL of sterile distilled water (autoclaved at 120 °C, 1 atm, 15 min). Dissolve with the aid of a glass stem with stirring up to complete homogenization. Label, identify, and store.
4. 0.1 M hydrochloric acid (HCl) solution. Check the molarity of HCl on the packaging label. Calculate the required volume of HCl to prepare 100 mL of 0.1 M solution. Dilute the volume of concentrated HCl in a glass container with 100 mL of sterile distilled water. Dissolve with the aid of a glass stem with stirring up to complete homogenization. Label, identify, and store.
5. 1% (v/v) aqueous 0.1 M acetic acid solution: Add 1 mL of glacial acetic acid for each 100 mL of sterile distilled water in a glass container. Adjust the pH to 5.6 with NaOH 3 M solution. Mix the solution with magnetic stirrer.
6. Magnetic stirrer.
7. Potentiometer with a combined glass electrode.
8. EO emulsions: Select the EO and final concentrations to be used in emulsion prepared with sterile distilled water (45 ± 2 °C). Add 0.5–1 mL Tween 80 (0.5–1% v/v) for each 100 mL of water as a stabilizing agent (*see* **Notes 5** and **6**).
9. Production of Chi and EO coatings: Glycerol. Add 2–2.5 mL for each 100 mL of the dispersion immediately after dissolving the Chi in aqueous acetic acid solution and add the EO in selected final concentration prepared previously (*see* **Note 7**).

2.3 Effects of Chi and EO on Fungal Spore Germination

1. Laboratory tools: automatic micropipettes, plastic 1.5 mL micro tubes, laboratory glassware, and sterile gauze.
2. 0.85% (w/v) sterile saline solution. Weigh 0.85 g of NaCl for each 100 mL of solution. Transfer to a glass container. Add distilled/deionized water. Mix with the aid of a glass stem up to complete dissolution. Autoclave (120 °C, 1 atm, 15 min). Properly identify, date, and store.
3. Hemocytometer or Neubauer counting chamber for counting spores or other suspended particles.
4. Optical microscope.
5. Lactophenol cotton blue solution. Add 20 mL of sterile distilled water, 20 mL of lactic acid, 40 mL of glycerol, and 20 mL of concentrated phenol in this sequence in a glass container and mix for 1 h. Weigh 0.05 g of trypan blue on a precision analytical balance and add to solution. Homogenize for 1 additional hour. Store the solution in a dark glass under refrigeration.

2.4 Effects of Chi and EO on Fungal Mycelial Growth

1. Laboratory tools: automatic micropipettes, laboratory glassware, cork borer, stainless steel tweezers, and Petri dish (90 mm diameter).
2. Use Chi dispersions and EO emulsions (individually or in combination). Add the solutions directly into sterile semisolid culture medium, mix up to complete dissolution, and pour into Petri dishes (*see* **Note 8**).
3. BOD incubator (biological oxygen demand) of high degree temperature accuracy.
4. Pachymeter (capacity 150 mm), calibrated at 20 ± 2 °C to measure the mycelial growth.

2.5 Effects of Chi and EO on Fungal Postharvest Disease in Fruit

1. Laboratory tools: automatic micropipettes, pipettes, sterilized needle, stainless steel tweezers, nylon membrane filters (or absorbent filter papers), and laboratory glassware.
2. Fruit and disinfection: select fruit before reaching full horticultural maturity or at a pre-established degree of maturity based on available literature (*see* **Notes 9** and **10**).
3. 1 M NaOH solution. Check the purity of NaOH (% w/w) on the packaging label. Calculate the mass of NaOH needed to prepare 100 mL of 1 M solution. Prepare as described in Subheading 2.2.
4. Sodium hypochlorite solution (150 ppm). In a container add 1.5 mL of sodium hypochlorite for each 1 L of potable water. To ensure the efficacy of chlorination, adjust the solution pH to 7.2 with 1 M NaOH.
5. Humid chamber: use polyethylene containers, paper towels, plastic bags, and sterile distilled water (autoclaved at 120 °C, 1 atm, 15 min).
6. Negative control group for antifungal activity. Add 2–2.5 mL of glycerol to each 100 mL of sterile distilled water. Adjust the pH to 5.6 with 3 M NaOH solution.
7. Positive control group for antifungal activity. Use synthetic antifungals pre-selected as standard for comparison: Dissolve selected synthetic antifungal in sterile distilled water to obtain the recommended manufacturer concentration.
8. Pachymeter (capacity 150 mm), calibrated at 20 ± 2 °C to determine lesion diameter.

3 Methods

3.1 Fungal Strains and Growth Conditions

Different fungal species can be studied as target microorganisms in antifungal assays. Select the fungal strains based on postharvest disease and fruit species evaluated (*see* **Note 11**). Stocks of tested fungal strains must be maintained in disks (5 mm diameter) of potato dextrose agar (or sabouraud agar) prepared as described in Subheading 2; stored in micro tubes with 1.5 mL of sterile saline solution (1% w/v); and kept at 5 °C in the dark. For antifungal assays, obtain work cultures grown in potato dextrose agar (or sabouraud agar) for 7 days at 25 ± 2 °C in a BOD incubator.

3.2 Preparation of Chi and EO Dispersions and Coatings

Chi dispersions: Weigh Chi in adjusted amounts to obtain the desired final concentrations (w/v). Selection of Chi concentrations used to formulate coatings should consider the in vitro antimicrobial efficacy and lowest concentration capable of providing high viscous dispersions with coating features that allow its application on fruit [5, 8]. For dissolution, add the polymer in aqueous acetic acid solution with constant stirring (120–150 rpm) using a magnetic bar at room temperature (25 ± 2 °C) for 18 to 24 h, depending on Chi concentration.

To ensure that found antifungal activity must be linked to Chi (and not of acetic acid), adjust the pH of Chi dispersions to 5.5 or 5.6 (referred as suitable pH for fungal growth) after the dispersion of polymer by adding 3 M NaOH or 0.1 M HCl (*see* **Note 12**).

EO emulsions: Add the EO in selected concentrations (v/v) in sterile distilled water (45 ± 2 °C) with Tween 80 (1% v/v) as a stabilizing agent. Keep under magnetic stirring (120–150 rpm) under room temperature (25 ± 2 °C) up to complete emulsion. Type and concentrations of EO used to formulate the edible coatings must be selected considering results of in vitro assays [5].

Production of Chi and EO coatings: For application of Chi and EO in combination, initially dilute the polymer in acetic acid (1%, v/v) with constant stirring (120–150 rpm) for 6 h at room temperature (25 ± 2 °C). Add EO emulsions (in different concentrations, v/v), directly into the Chi dispersion, followed by stirring for additional 18 h at room temperature (25 ± 2 °C). For application of Chi and EO dispersion on fruit, add glycerol (2–2.5%, v/v) as a plasticizer, immediately after incorporation of EO into coating-forming dispersions. 0.5–1% (v/v) Tween 80 should be also used to facilitate homogenization of EO into Chi dispersions (*see* **Note 13**).

3.3 Effects of Chi and EO on Fungal Spore Germination

Collect fungal spores with addition of sterile saline solution (0.85% w/v) in a 7-day-old culture (25 ± 2 °C). Filtrate the resulting suspension with a multiple layer of sterile gauze to retain hyphal fragments. Using a hemocytometer in optical microscope (using magnification ×40), quantify the number of spores in suspension and adjust the concentration with sterile saline solution to yield an inoculum of approximately 10^6 spores/mL. Distribute an 0.1 mL aliquot of each spore suspension in micro tubes with 0.1 mL of Chi and EO dispersions individually or in combination (in selected concentrations). Subsequently, put 0.1 mL of the mixture in center

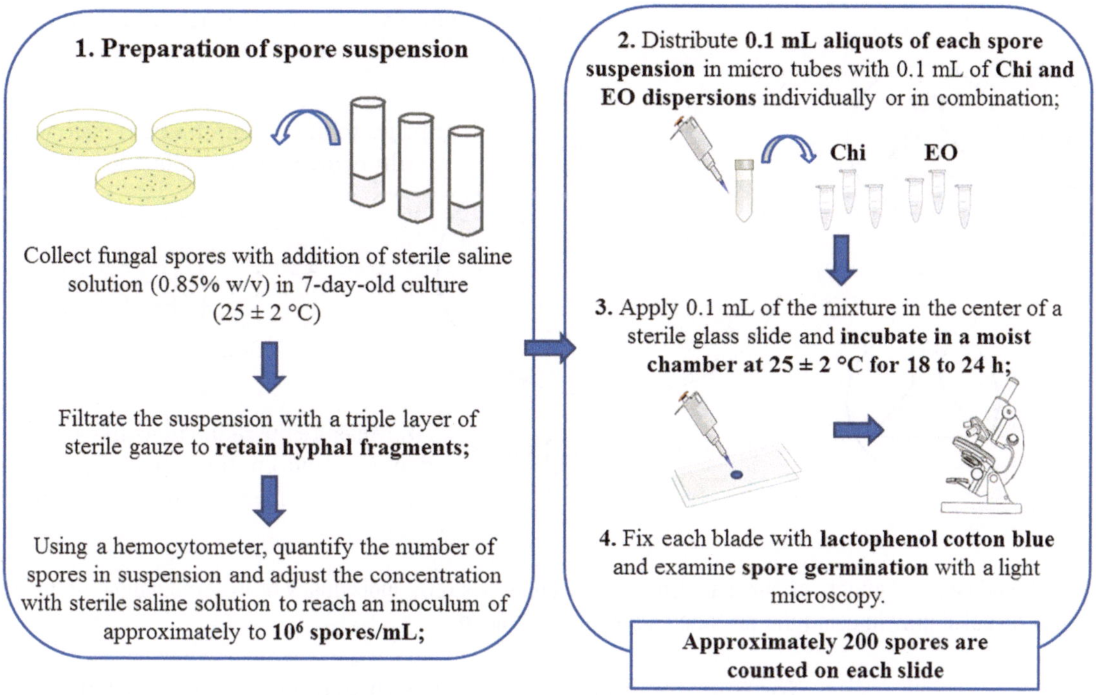

Fig. 1 Evaluation of effects of chitosan (Chi) and essential oils (EO), individually or in combination, on fungal spore germination

of a sterile glass slide and incubate in a moist chamber (relative humidity 85%) at 25 ± 2 °C for 18–24 h. Fix each blade with lactophenol cotton blue stain and examine spore germination with light microscopy (using magnification ×40) (Fig. 1). Approximately 200 spores are counted on each slide. As controls, similarly test suspensions of fungal spores cultivated in growth media without Chi and/or EO. Results are expressed as the percent of spore germination inhibition after comparing the number of germinated spores in media with Chi and EO with that found in control treatment [12, 13] (*see* **Note 14**).

3.4 Effects of Chi and EO on Fungal Mycelial Growth

Effects of Chi and EO (individually or in combination) on radial mycelial growth are assessed with poisoned substrate technique (dilution in solid media), as described in Fig. 2.

Initially, with the aid of a sterile cork borer, remove mycelial agar plugs (5 mm diameter) from margin of fungal culture grown for 7 days at 25 ± 2 °C. Transfer with the aid of a sterilized stainless steel tweezer to the center of a Petri dish with selected culture medium with Chi and/or EO at the desired final concentration and incubated during 10 days at 25 ± 2 °C. Test culture medium without Chi or EO (pH 5.6) as negative control (*see* **Note 15**). Daily measure the radial mycelial growth for 10 days or when the negative control Petri dish was fully completely with mycelia.

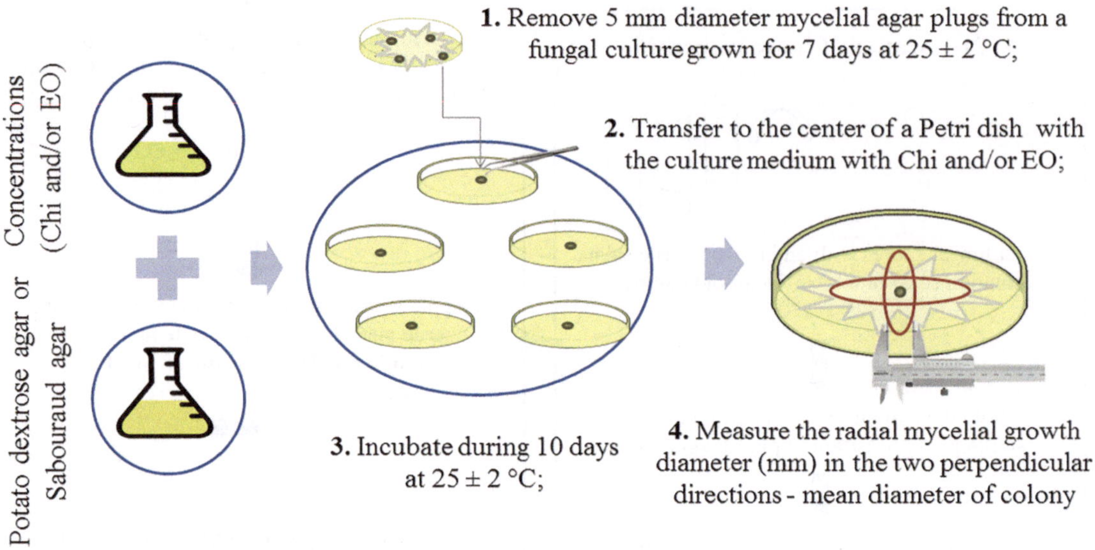

Fig. 2 Evaluation of effects of chitosan (Chi) and essential oils (EO), individually or in combination, on fungal radial mycelial growth using the poisoned substrate technique

Measure the radial mycelial growth diameter (mm) of each fungal colony in two perpendicular directions with a pachymeter to get mean diameter of colony. Calculate the percentage of mycelial growth inhibition (MGI %) with the formula: MGI $\% = [(C - T)/C] \times 100$, where C is the colony diameter in control assay and T is the diameter of colony growing in selected culture medium with Chi and/or EO [7, 14].

3.5 Evaluation of the Type of Interaction of Chi and EO

Determine the type of interaction of Chi and EO combinations with Abbott method [15, 16]. For this, consider the MGI% caused by Chi or EO individually and in combination, being named of MGI% observed ($MGI\%_{obs}$). Use the poisoned substrate technique described previously to obtain these results. In sequence, calculate the MGI% expected ($MGI\%_{exp}$) for each combination with the formula: $MGI\%_{exp} = MGIChi\%_{obs} + MGIEO_{obs}\% - [(MGIChi\% \times MGIEO\%)/100]$, where $MGIChi\%_{obs}$ and $MGIEO\%_{obs}$ are the individual values of $MGI\%_{obs}$ for Chi and EO at the tested concentrations, respectively.

Finally, determine the effects of Chi and EO combinations calculating the Abbott index (AI), as follows: $AI = MGI\%_{obs} / MGI\%_{exp}$, where $MGI\%_{obs}$ is the result found with application of the combination of Chi and EO. Results are interpreted as synergistic effect for AI >1.5, additive effect for AI ≥ 0.5 to 1.5, and antagonistic effect for AI <0.5.

3.6 Effects of Chi and EO on Fungal Postharvest Diseases in Fruit

Disease incidence (percentage of fruit showing characteristic symptoms of a specific disease) and disease severity (lesion development) can be measured in fruit by inoculation of selected fungal strain (spore suspension or agar plug with mycelia) [3, 5, 7]. Curative effects are assessed by inoculation of fungal strain, followed by coating the fruit, while the preventive effects are assessed by coating the fruit, followed by inoculation of fungal strain (Fig. 3).

Prior to assays, wash the fruit in running water, disinfect the surface by immersing in a sodium hypochlorite solution (150 ppm) for 5 min, wash again with sterile distilled water, and dry for 2 h in a safety cabinet [3, 7].

For assays to measure preventive effects (and curative effects), make injuries in fruit surface (epidermis, five wounds per fruit) with a sterilized needle (3 mm depth and 2 mm wide) in middle fruit region on the same side [3, 4, 9] (*see* **Note 16**).

Immerse each fruit carefully in a glass container with 500 mL of coating-forming dispersions formulated with combinations of selected Chi and EO concentrations for a time interval of 1–5 min [3, 4, 7] (*see* **Note 17**). Then, completely air-dry the fruit on a nylon filter (or absorbent filter papers) to drain the excess liquid in a biosafety cabinet for 2 h. Subsequently, artificially inoculate the fruit:

1. Via immersion in 500 mL of an inoculum solution (approximately 10^6 spores/mL) of the tested fungal species for 1 min, with gentle stirring using a sterile glass stem, and left to dry in a biosafety cabinet for 1 h (25 ± 2 °C) [12, 13].
2. Agar plug (5 mm diameter) with mycelia obtained from margin of a 7-day-old colony grown on selected culture medium at 25 ± 2 °C and transferred with the aid of a sterilized stainless steel tweezers to each injury/puncture point in fruit [3, 7] (*see* **Notes 18** and **19**).

For assays focusing on curative effects: Inoculate the fruit via immersion in 500 mL of a spore solution (approximately 10^6 spores/mL) of selected fungal strain or with an agar plug of mycelia (in puncture point on fruit surface). Afterwards, perform the immersion of inoculated fruit (and non-inoculated controls) in coating-forming dispersion with Chi and EO for 1–5 min and completely air-dry the fruit on a nylon filter to drain excess liquid in a biosafety cabinet for 2 h [7].

Subsequently, for both assays, distribute the fruit individually in polystyrene trays lined with paper towels previously moistened with sterile distilled water (forming a humid chamber) and covered with a commercially sterile plastic bag to maintain a high relative humidity (*see* **Note 20**). Keep the containers at room temperature (25 ± 2 °C) and remove the humidity chambers after 48 h [7, 17].

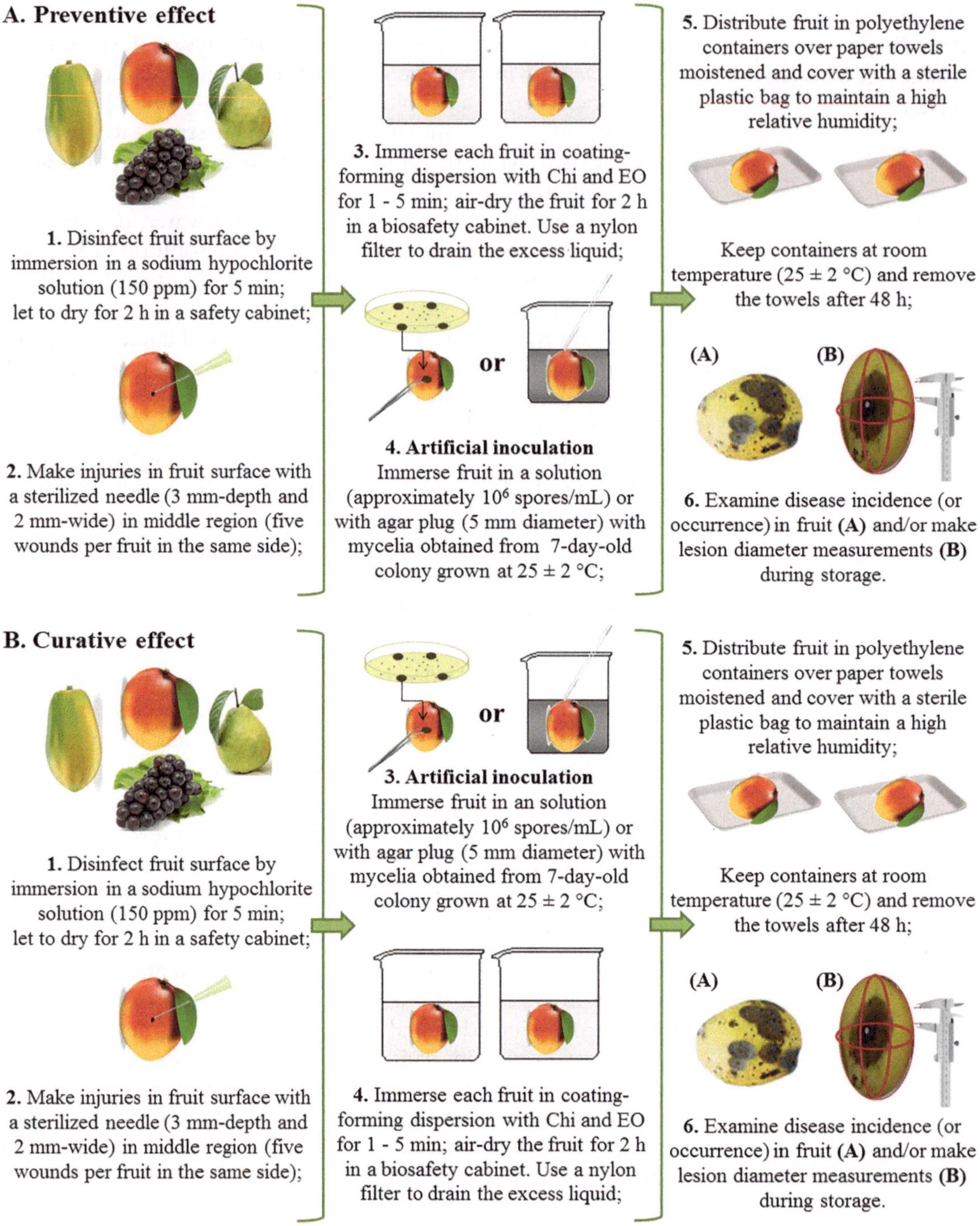

Fig. 3 Description of preventive and curative assays for assessing the effects of formulated coatings with of chitosan (Chi) and essential oils (EO) when applied on fruit. (**a**) Preventive effect; (**b**) curative effect

A negative control using sterile distilled water and glycerol replacing the coating-forming dispersion should be done. For positive controls, use commercial formulation of synthetic fungicides commonly applied to inhibit target fungal species (e.g., thiophanate-methyl; difenoconazole; trifloxystrobin; tebuconazole) [3, 4, 7].

To avoid external contamination during storage, the experiments to examine the visible characteristic symptoms of fungal infection and/or lesion diameter measurements must be done in a safety cabinet. Number of fruit is defined based on a completely randomized experimental design, and all fruit should be kept in experiment up to the end of evaluated storage period.

Results of the effects of Chi and EO on disease incidence (or occurrence) in fruit are expressed as the storage time needed for appearance of visible signs of fungal infection and as the percentage of infected fruit at assessed different storage time intervals [12, 13]. Effects of treatments on disease severity in fruit are measured based on percentage of lesion diameter reduction (% LDR), determined by difference in lesion diameter in fruit coated with Chi and EO combination or treated with fungicide formulation (positive control) compared to lesion diameter in uncoated/ untreated fruit (negative control) [3]. Measure lesion diameter (mm) in fruit in two perpendicular directions with a pachymeter to get mean diameter.

Calculate %LDR with the formula: $\%LDR = [(N - F/N] \times 100$, where N is the lesion diameter in negative control fruit and F is the lesion diameter in fruit coated with Chi and EO combinations or treated with synthetic fungicides [3, 4, 8].

4 Notes

1. Fungi are the most important causal agents of fruit postharvest diseases worldwide [5]. Some of the main fungi studied as causative agents of postharvest diseases in tropical fruit are *Aspergillus flavus*; *Aspergillus niger*; *Botrytis cinerea*; *Colletotrichum asianum*; *Colletotrichum siamense*; *Colletotrichum fructicola*; *Colletotrichum karstii*; *Colletotrichum tropicale*; *Colletotrichum gloeosporioides*; *Colletotrichum brevisporum*; *Penicillium digitatum*; *Penicillium expansum*; and *Rhizopus stolonifer*, mostly obtained from collections of research institutes [4, 7, 8, 13, 17–20].
2. Culture medium: It is important to select a culture medium suitable to provide all nutrients required for growth of selected fungal strain. Most common media used for fungal growth are potato dextrose agar and sabouraud agar.
3. Molecular weight and deacetylation degree are factors that may influence the bioactivities of Chi. It is recommended to use Chi of medium molecular weight and degree of deacetylation of $\geq$75% [5].

4. Commercial crustacean Chi is often used. However, Chi extracted from fungi (class *Zygomycetes*, order *Mucorales*) and shrimp shells may have characteristic of medium molecular weight and/or degree of deacetylation of $\geq$75% [8, 13, 18].
5. Chemical composition of EO needs to be identified because the antifungal effects depend on the proportion of individual bioactive components present in EO [21]. Some EO with recognized antifungal activity already assayed with methods described here are *Mentha* (*Mentha piperita* L.; *Mentha* × *villosa* Huds.); lemongrass (*Cymbopogon citratus* (D.C. ex Nees) Stapf.); oregano (*Origanum vulgare* L.); clove (*Syzygium aromaticum* (L.) Merr.); and rosemary (*Rosmarinus officinalis* L.) [3, 4, 8, 12, 20, 22].
6. Addition of stabilizing agent (Tween) may facilitate in adhesion of coatings to fruit surfaces and stabilize coating matrices, preventing lipid migration, by reducing the surface tension of water-lipid interfaces [6].
7. Plasticizers such as glycerol are often added to coatings to increase flexibility and prevent blistering, flaking, and cracking [6].
8. Use a 70–75% proportion of added growth medium and 25–30% of the antimicrobial solution, with concentrations adjusted to the total volume. For example, a total volume of 400 mL: 297.5 mL of PDA (or sabouraud agar) added to 100 mL acetic acid with 2000 mg Chi (concentration 5 mg/mL) + 2 mL Tween 80 and 0.5 mL EO (concentration 1.25 μL/mL).
9. Different fruits are susceptible to the action of phytopathogens and have already been investigated for the efficacy of the use of Chi-EO coatings using the methods described here, including table grapes, mango, papaya, guava, mandarins, and banana [3, 4, 7, 8, 12, 13, 20].
10. Some requirements must be taken into account to obtain these fruit, such as uniformity of shape, size, and maturity (based on fruit peel color). Fruit must be free from injuries or infections by insects or pathogens.
11. It is important that all tested fungal strains are capable of inducing the development of symptoms/injuries characteristic of the diseases in examined fruit, so that they can be re-isolated and used in in situ experiments.
12. Slowly add the reagents to prevent Chi precipitation.
13. Vigorous stirring can be applied for homogenization of low concentrations (<1 μL/mL) of EO in Chi dispersion, without adding Tween 80.

14. A spore (conidium) is considered germinated when the developed germ tube was at least twice its original diameter [13].
15. Culture medium with aqueous 0.1 M acetic acid (1% v/v) solution (pH 5.6) and Tween 80 (final concentration 0.5–1% v/v) should be also tested as a control and should not have inhibitory effects on radial mycelial growth of tested fungal isolates [4, 7].
16. Injuries in fruit surface are necessary because most phytopathogenic fungi invade and cause lesion via wounds or microcracks in fruit peel [17].
17. Time for immersion in fruit coating dispersions will vary according to fruit size, and entire fruit surface must be uniformly coated.
18. The agar plug must be fixed with the aid of a tape on fruit surface [3, 4].
19. Fruit should be inoculated with colonized agar plug method when tested fungal strain does not produce conidia in culture [17, 23].
20. Glass plates are used as support to avoid contact of fruit with paper towels.

Acknowledgments

The authors thank the National Council for Scientific and Technological Development—CNPq (Brazil) (grant numbers 403122/2016-3 and 408724/2018-8) and Coordination of Higher Personnel Improvement—CAPES (Brazil) (Finance code 001) for funding and supporting these researches.

References

1. de Oliveira KÁR, da Conceição ML, de Oliveira SPA et al (2019) Postharvest quality improvements in *mango* cultivar Tommy Atkins by chitosan coating with *Mentha piperita* L. essential oil. J Hortic Sci Biotechnol 95:260–272. https://doi.org/10.1080/14620316.2019.1664338
2. Hosseini SF, Amraie M, Salehi M et al (2019) Effect of chitosan-based coatings enriched with savory and/or tarragon essential oils on postharvest maintenance of kumquat (*Fortunella* sp.) fruit. Food Sci Nutr 7:155–162. https://doi.org/10.1002/fsn3.835
3. Oliveira PDL, de Oliveira KÁR, dos Santos Vieira WA et al (2018) Control of anthracnose caused by *Colletotrichum* species in guava, mango and papaya using synergistic combinations of chitosan and *Cymbopogon citratus* (D.C. ex Nees) Stapf. essential oil. Int J Food Microbiol 266:87–94. https://doi.org/10.1016/j.ijfoodmicro.2017.11.018
4. Braga SP, Lundgren GA, Macedo SA et al (2019) Application of coatings formed by chitosan and *Mentha* essential oils to control anthracnose caused by *Colletotrichum gloesporioides* and *C. brevisporum* in papaya (*Carica papaya* L.) fruit. Int J Biol Macromol 139:631–639. https://doi.org/10.1016/j.ijbiomac.2019.08.010

5. de Souza EL, Lundgren GA, de Oliveira KÁR et al (2019) An analysis of the published literature on the effects of edible coatings formed by polysaccharides and essential oils on postharvest microbial control and overall quality of fruit. Compr Rev Food Sci Food Saf 18: 1947–1967. https://doi.org/10.1111/1541-4337.12498
6. Riva SC, Opara UO, Fawole OA (2020) Recent developments on postharvest application of edible coatings on stone fruit: a review. Sci Hortic (Amsterdam) 262:109074. https://doi.org/10.1016/j.scienta.2019.109074
7. Oliveira KÁR, Berger LRR, de Araújo SA et al (2017) Synergistic mixtures of chitosan and *Mentha piperita* L. essential oil to inhibit *Colletotrichum* species and anthracnose development in mango cultivar Tommy Atkins. Food Microbiol 66:96–103. https://doi.org/10.1016/j.fm.2017.04.012
8. Guerra ICD, De Oliveira PDL, Santos MMF et al (2016) The effects of composite coatings containing chitosan and Mentha (*piperita* L. or x *villosa* Huds) essential oil on postharvest mold occurrence and quality of table grape cv. Isabella. Innov Food Sci Emerg Technol 34:112–121. https://doi.org/10.1016/j.ifset.2016.01.008
9. Gomes ACA, da Costa LM, de Oliveira KÁR et al (2020) Coatings with chitosan and phenolic-rich extract from acerola (*Malpighia emarginata* D.C.) or jabuticaba (*Plinia jaboticaba* (Vell.) Berg) processing by-product to control rot caused by *Lasiodiplodia* spp. in papaya (*Carica papaya* L.) fruit. Int J Food Microbiol 331:108694. https://doi.org/10.1016/j.ijfoodmicro.2020.108694
10. Nair MS, Saxena A, Kaur C (2018) Effect of chitosan and alginate based coatings enriched with pomegranate peel extract to extend the postharvest quality of guava (*Psidium guajava* L.). Food Chem 240:245–252. https://doi.org/10.1016/j.foodchem.2017.07.122
11. Tayel AA, Gharieb MM, Zaki HR, Elguindy NM (2016) Bio-clarification of water from heavy metals and microbial effluence using fungal chitosan. Int J Biol Macromol 83:277–281. https://doi.org/10.1016/j.ijbiomac.2015.11.072
12. Barreto TA, Andrade SCA, Maciel JF et al (2016) A chitosan coating containing essential oil from *Origanum vulgare* L. to control postharvest mold infections and keep the quality of cherry tomato fruit. Front Microbiol 7:1724. https://doi.org/10.3389/fmicb.2016.01724
13. Guerra ICD, de Oliveira PDL, de Souza Pontes AL et al (2015) Coatings comprising chitosan and *Mentha piperita* L. or *Mentha*×*villosa* Huds essential oils to prevent common postharvest mold infections and maintain the quality of cherry tomato fruit. Int J Food Microbiol 214:168–178. https://doi.org/10.1016/j.ijfoodmicro.2015.08.009
14. Camiletti BX, Asensio CM, de la Paz Gimenez Pecci M, Lucini EI (2014) Natural control of corn postharvest fungi *Aspergillus flavus* and *Penicillium* sp. using essential oils from plants grown in Argentina. J Food Sci 79: M2499–M2506. https://doi.org/10.1111/1750-3841.12700
15. Camiletti BX, Asensio CM, Gadban LC et al (2016) Essential oils and their combinations with iprodione fungicide as potential antifungal agents against withe rot (*Sclerotium cepivorum* Berk) in garlic (*Allium sativum* L.) crops. Ind Crop Prod 85:117–124. https://doi.org/10.1016/j.indcrop.2016.02.053
16. Kosman E, Cohen Y (1996) Procedures for calculating and differentiating synergism and antagonism in action of fungicide mixtures. Phytopathology 86:1263–1272
17. Lima NB, Lima WG, Tovar-Pedraza JM et al (2015) Comparative epidemiology of *Colletotrichum* species from mango in northeastern Brazil. Eur J Plant Pathol 141:679–688. https://doi.org/10.1007/s10658-014-0570-y
18. Oliveira CEV, Magnani M, de Sales CV et al (2014) Effects of chitosan from *Cunninghamella elegans* on virulence of post-harvest pathogenic fungi in table grapes (*Vitis labrusca* L.). Int J Food Microbiol 171:54–61. https://doi.org/10.1016/j.ijfoodmicro.2013.11.006
19. Athayde AJAA, de Oliveira PDL, Guerra ICD et al (2016) A coating composed of chitosan and *Cymbopogon citratus* (Dc. Ex Nees) essential oil to control Rhizopus soft rot and quality in tomato fruit stored at room temperature. J Hortic Sci Biotechnol 91:582–591. https://doi.org/10.1080/14620316.2016.1193428
20. Shao X, Cao B, Xu F et al (2015) Effect of postharvest application of chitosan combined with clove oil against citrus green mold. Postharv Biol Technol 99:37–43. https://doi.org/10.1016/j.postharvbio.2014.07.014
21. Taghavi T, Kim C, Rahemi A (2018) Role of natural volatiles and essential oils in extending shelf life and controlling postharvest microorganisms of small fruits. Microorganisms 6: 1–15. https://doi.org/10.3390/microorganisms6040104

22. De Sousa LL, de Andrade SCA, Athayde AJAA (2013) Efficacy of *Origanum vulgare* L. and *Rosmarinus officinalis* L. essential oils in combination to control postharvest pathogenic Aspergilli and autochthonous mycoflora in *Vitis labrusca* L. (table grapes). Int J Food Microbiol 165:312–318. https://doi.org/10.1016/j.ijfoodmicro.2013.06.001

23. Vieira WAS, Michereff SJ, de Morais MA et al (2014) Endophytic species of *Colletotrichum* associated with mango in northeastern Brazil. Fungal Divers 67:181–202. https://doi.org/10.1007/s13225-014-0293-6

Chapter 12

Ohmic Heating for Food Processing: Methods and Procedures Related to Process Parameters

Pitiya Kamonpatana, Mohsen Gavahian, and Sudhir K. Sastry

Abstract

Ohmic heating has a great promise of producing safe food products while maintaining their quality, since volumetric, uniform, and rapid heating make it superior to conventional heat treatment. Ohmic heating facilities have been available worldwide and have processed a wide range of foods. Electrical conductivity and electric field strength are among the key factors of heating performance. For solid-liquid food mixtures, the solid and liquid phases can be heated at the same rate when their electrical conductivity is in the same range. This chapter describes some factors (*e.g.*, ionic content, solid structure, and food formulation) that affect electrical conductivity and heat generation in ohmic treatments. We also outline herein procedures of color change analysis, microbial assay, and vitamin C degradation as examples of typical measures to monitor the safety, performance, and quality parameters in an ohmic process to provide fundamental procedures for new researchers and students in this field.

Key words Electrical conductivity, Inoculated pack, Ohmic heating, Solid-liquid mixture, Structure, Vitamin C, Protocols

1 Introduction

Ohmic heating is an emerging thermal processing technology, allowing alternating current to pass through food products which converts electrical energy into heat. The advantage of ohmic heating over conventional heating is its rapidity and uniformity, resulting in reduced heat damage and nutrient loss. The heating rate of ohmic heating depends on the electrical conductivity of the material and the electric field strength. Procedures to reduce the risk of under-processing and improve food quality include (1) determination of electrical conductivity, (2) identification of the slowest heating point, and (3) verification of the velocity of the fastest component [1]. The cellular structure of solid foods, including fruits, vegetables, and muscle foods, affects electrical conductivity [2]. Electrical conductivity of intact and damaged cellular materials shows differences [2, 3]. Meat with fiber oriented parallel to the

Mohsen Gavahian (ed.), *Emerging Food Processing Technologies*,
Methods and Protocols in Food Science, https://doi.org/10.1007/978-1-0716-2136-3_12,

electric field has electrical conductivity higher than that perpendicular to the electric field [2, 4]. Ionic components play an essential role as increasing salt, and acid content increases electrical conductivity. Uniform heat generation and temperature distribution are preferable for foods that are solid-liquid mixtures. However, food components with different electrical conductivities may result in nonuniform heating. Food formulation of solid particles using soaking, blanching, and vacuum in water or salt solutions of different concentrations can adjust electrical conductivity to be comparable to the liquid phase [5, 6]. The formulation method may therefore be used to improve temperature distribution.

Ohmic heating may offer less heat damage and better retention of nutrients, vitamins, pigments, and aromas [7–9]. Vikram et al. [10] revealed that ohmic heating maintained the highest values of vitamin C and carotenoids in orange juice compared with infrared, microwave, and conventional heating. Jaeschke et al. [11] reported the same degradation of ascorbic acid and carotenoids in acerola pulp between ohmic and conventional heating. On the other hand, higher degradation of total vitamin C in acerola pulp was observed at ohmic heating with high electric field compared to conventional heating [12]. Degradation processes under ohmic heating may depend on field strength, frequency, and temperature [11]. It is necessary to investigate the effect of ohmic heating conditions on food properties and bioactive compound degradation.

We outline herein a set of experiments to enable beginning researchers to become familiar with ohmic heating technology. While the actual operational details for a commercial heater could be very different (continuous flow vs. batch), experience with a batch system could be useful in understanding the technology. Users are cautioned that results for a commercial or large-scale system with real foods and high solids concentration may be very different from what is observed in these cases. We also emphatically caution that there is no substitute for a deep-level understanding of the subject; hence this is only a guide to get started.

Here we outline a series of steps to set up a static ohmic heating system and demonstrate the effect of salt content and structure on electrical conductivity and for measurement of temperature profiles of solid-liquid mixtures with particulates of electrical conductivity higher or lower than the liquid medium. A comparison of orange juice treated with ohmic and conventional heating on pH, color, and microbial count is detailed. In addition, the technique of vitamin C determination is also discussed.

2 Materials and Methods

2.1 Set Up a Static Ohmic Heating System (Fig. 1) (See Notes 1–3)

1. Equip static ohmic heating cell (*see* details in **Note 4**) with two electrodes (Fig. 2) (*see* details in **Note 5**).
2. Hold the ohmic heating cell in place using a holding stand.
3. Wire the power cable from a power supply (*see* **Note 6**) to the electrodes.
4. Connect the volt transducers across two cables linking to two electrodes and current transducers to one cable.
5. Insert thermocouples (*see* **Notes 7–9**) at the geometric center of the ohmic heating cell (and other positions).
6. Wire the cables from voltage and current transducers and thermocouple to the data acquisition connecting to a personal computer.
7. Monitor and record temperature, current, and voltage with the data acquisition device (*see* **Note 9**), which transfers the information to a computer. These types of data are necessary for process evaluation such as energy calculation, heating rate, come-up time, etc.

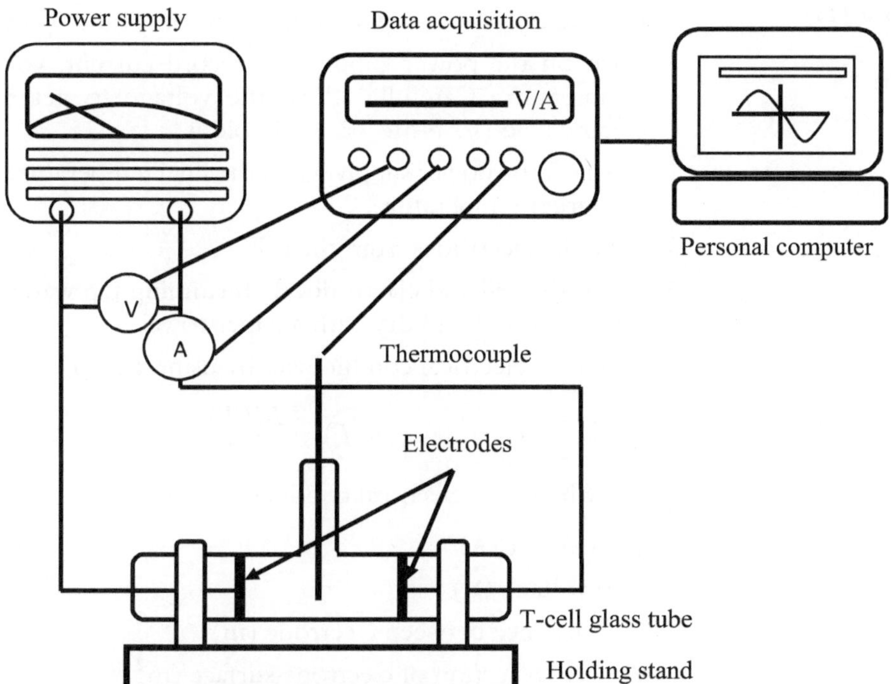

Fig. 1 Schematic diagram of static ohmic heating system

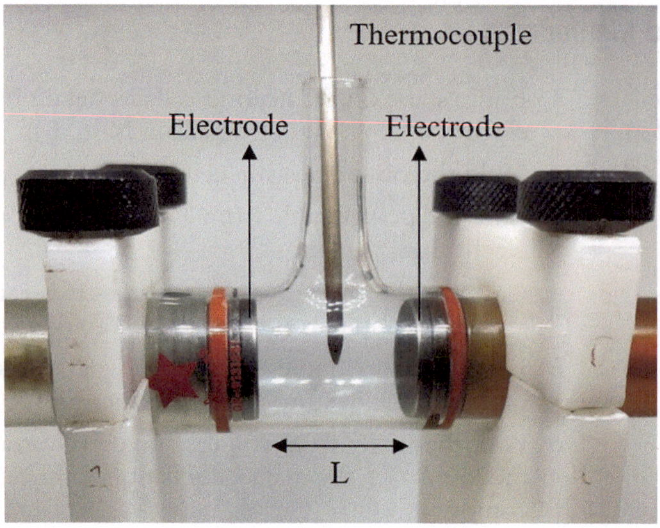

Fig. 2 Setup of ohmic heating cell and electrodes

2.2 Electrical Conductivity Measurement

Different concentrations of various types of salts (e.g., NaCl) can be used. Here, as two examples, salt solutions of 0.2% and 0.5% (w/v) sodium sulfate will be used to provide a protocol.

2.2.1 Electrical Conductivity Measurement of Liquid Foods (See Notes 10 and 11)

Salt Solution

1. Add 0.2% (w/v) salt solution to the ohmic heating cell through the top opening.
2. Insert thermocouple at the geometric center of the ohmic heating cell. Avoid any air bubbles in the liquid foods.
3. Turn on the power supply and record current, voltage, and temperature. Carefully adjust the voltage to get the target electric field (*see* **Note 12**) and hold it.
4. Turn off the power supply when the final temperature arrives at the target temperature.
5. Take the electrodes from the cell.
6. Clean the cell and electrodes with running tap water and delicate detergent and dry with a paper towel.
7. Calculate electrical conductivity by using Eq. (1):

$$\sigma(T) = \frac{LI(T)}{VA} \tag{1}$$

where σ = electrical conductivity (S/m)

I = current (A)

V = voltage (V)

L = distance between electrode (m)

A = cross section of electrode surface (m^2)

8. Fill obtained data in Table 1. Spreadsheets can be used to facilitate data input and analysis.

Table 1
An example approach to organize some of the major parameters in ohmic processing

Time (s)	Temperature (C)	Voltage (V)	Current (A)	Electrical conductivity (S/m)

9. Plot the electrical conductivity vs. temperature.
10. Change to 0.5% (w/v) salt solution and repeat the step.
11. Apply the same electric field and compare the result.

Prepare all solutions using deionized water and analytical grade reagents. If the sample has sufficient electrical conductivity to allow for heating, the process can be performed without addition of electrolytes such as salt. It is important to understand the extent of statistical variability in the sample prior to all studies. Performing all procedures in triplicate is suggested.

*2.2.2 Electrical Conductivity Measurement of Solid Foods (See **Notes 10** and **11**)*

1. Place the solid piece in the ohmic heating cell and equip it tightly with two electrodes (*see* **Notes 11** and **13**).
2. Measure the distance between electrodes.
3. Insert thermocouple at the geometric center of the ohmic heating cell. Ensure the same positions for every run.
4. Set up and turn on the system. Apply voltage to get the target electric field and heat the sample to the target temperature.
5. Record temperature, voltage, and current.
6. Calculate electrical conductivity by using Eq. (1) where A is the cross section of solid surface (m^2) and fill obtained data in the spreadsheet mentioned in Subheading "Salt Solution".

Comparison of Electrical Conductivity of Intact and Damaged Plant Cells

Electrical conductivity of solid foods depends on their cellular structure. Intact and damaged plant cells showed the difference in electrical conductivity behavior. Keep fresh carrots to represent intact cells and blanched carrots as damaged cells and compare their electrical conductivity as following procedure.

Fresh and Blanched Carrots

1. Peel carrots and cut them into pieces.
2. To conduct blanched carrots, seal carrots in an airtight plastic bag and immerse in boiling water for 6 min. Leave them cool.
3. Comparison of the electrical conductivity of the intact and damaged cells (from Subheading "Salt Solution").

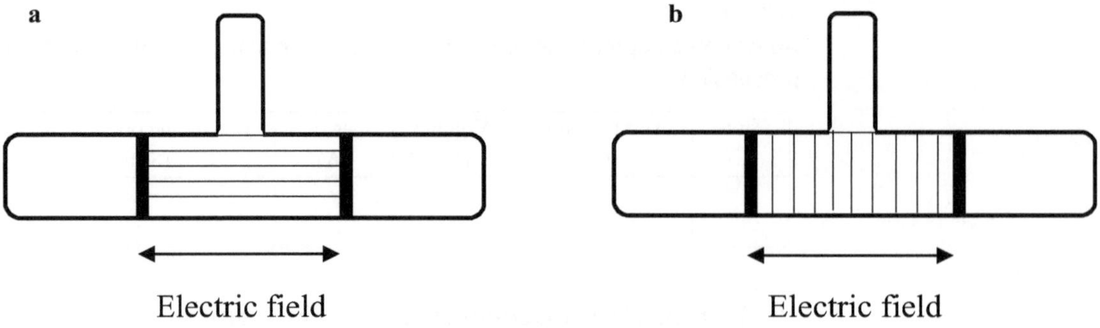

Fig. 3 Meat with fiber (**a**) oriented parallel and (**b**) perpendicular to electric field

Comparison of Electrical Conductivity of Plant Cells with Different Pretreatments

To perform heating uniformity of the solid-liquid food mixtures, their electrical conductivity should be matched as closely as possible. Pretreatments such as boiling in water or salt solution have been introduced to alter the electrical conductivity.

Comparison of Electrical Conductivity of Carrots Blanched in Salt Solution and Water

1. Put carrots in a boiling solution of 3.0% (w/v) sodium chloride for 6 min and leave them cool.
2. Comparison of the electrical conductivity of the carrots blanched in salt solution and water.

Comparison of Electrical Conductivity of Meat with Fiber Oriented Parallel and Perpendicular to the Electric Field

1. Meat with the uniformity of fiber orientation should be chosen.
2. Place the sample with its fiber in parallel orientation to electric field (Fig. 3a) and measure the electrical conductivity.
3. Place the sample with its fiber in perpendicular orientation to electric field (Fig. 3b) and repeat the step.
4. Comparison of their electrical conductivity.

2.3 Temperature Profiles of Liquid Foods Containing a Single Particle Under Static Ohmic Heating System (See Note 14)

*2.3.1 Particle Less Conductive Than the Medium (See **Note 15**)*

1. Insert thermocouple probe to the center of the carrot blanched in water for 6 min and set its position at the chamber's center. Locate two thermocouple probes in the front and side 5 mm from the particle.
2. Fill 0.2% (w/v) salt solution.
3. Turn on the system and record temperature, voltage, and current.
4. Apply voltage to get the target electric field and heat the carrot to the target temperature.
5. Fill obtained data in the Table 2.

*2.3.2 Particle More Conductive Than the Medium (See **Note 15**)*

1. Change to the carrot blanched in 3.0% (w/v) salt solution for 6 min and repeat the step in Subheading 2.3.1.

Table 2
An example method of organizing temperature data in an ohmic treatment to obtain temperature profile

		Temperature (°C)		
		Carrot		
Solid-liquid food mixture	**Time (s)**	**Center**	**Front**	**Side**
Particulate with less conductive than the medium				
Particulate with more conductive than the medium				

2.3.3 Comparison on Temperature Profiles

1. Plot temperatures vs. time of particle less conductive compared with particle more conductive than the medium.
2. To perform the agitated (mixed) condition, repeat the procedure with another setup and place the ohmic chamber on a shaker table.
3. To investigate the effect of a higher solids concentration, repeat the procedure with adding more solid pieces or change the solid to a larger size.

2.4 Orange Juice Treated Between Static Ohmic Heating and Conventional Heating (See Note 16)

2.4.1 Orange Juice

Clean and squeeze orange fruits to obtain orange juice. Filter the orange juice through a filter with 1 mm diameter holes to remove pulp and large particles.

2.4.2 Conventional Heating

1. Add fresh orange juice to a flask or beaker and heat with an electrical heater/hot plate (or water bath). Cover the opening with foil and insert a thermocouple at the geometric center of the beaker. Turn on the heater and magnetic stirrer. Record temperature and current (*see* **Note 17**). Heat the orange juice to 90 °C and hold for 12 s (recommendation). After treatment, cool the orange juice and store it at −20 C until further investigation.

2.4.3 Ohmic Heating

1. Fill fresh orange juice into the static ohmic heating cell (*see* **Note 16**) with the same amount as conventional heating.
2. Insert thermocouple at the geometric center of the ohmic heating cell. Avoid any air bubbles in orange juice.
3. Turn on the power supply and record current, voltage, and temperature. Slowly adjust the voltage to get the target electric field.

4. Turn off the power supply when the final temperature arrives at the target temperature of 90 °C with the holding time of 12 s (recommendation).
5. Cool the orange juice and store it at −20 °C until further investigation.
6. Take off the electrodes.
7. Clean the ohmic heating cell and electrodes with running tap water and delicate detergent and dry with a paper towel.
8. Plot temperature vs. time and compare temperature profile of orange juice treated by ohmic heating with conventional heating.

2.4.4 Product Analysis

Compare product properties (pH, total soluble solid content, and color), microbial counts (total plate counts and yeast and mold counts), and energy consumption of orange juice treated by ohmic heating with conventional heating.

1. pH: Measure pH using a pH meter (*see* **Note 18**).
2. Total soluble solid content: Detect total soluble solid content with a refractometer. Set zero set before use.
3. Color: Determine color using the CIE L^*, a^*, and b^* system and a spectrophotometer (*see* **Note 19**). The total color difference is calculated as

$$\Delta E = \left(\Delta L*^2 + \Delta a*^2 + \Delta b*^2\right)^{1/2} \tag{2}$$

where $\Delta L^* = L^* - L^*_0, \Delta a^* = a^* - a^*_0,$ and $\Delta b^* = b^* - b^*_0$. The subscript "0" indicates an untreated or fresh sample.

2.4.5 Microbiological Experiment (See **Notes 20–22**)

Determine Total Plate Counts Using the Pour Plate Method (Adapted from [13])

1. Serially dilute orange juice with 0.1% sterile peptone water.
2. Place 1.0 mL of the appropriate dilution into each Petri plate and add 15–25 mL of liquid plate count agar.
3. Leave the media to solidify and invert the plate.
4. Incubate plate at 35 °C for 48 h.
5. Perform each test in duplicate. Express the results as colony-forming units per milliliter (CFU/mL).

Determine Yeast and Mold Counts Using the Spread Plate Method (Adapted from [14])

1. Serially dilute orange juice with 0.1% sterile peptone water.
2. Add 0.1 mL of the appropriate dilution onto solidify dichloran rose Bengal chloramphenicol agar (DRBC).
3. Spread suspension evenly on medium surface with a sterilized bend glass rod.

Table 3
A sample table for heat generation and energy consumption calculation in ohmic studies

t (s)	Δ*t* (s)	Voltage (V)	Current (A)	Electric power (kW)	Energy consumption (kWh)
t_0					
t_1	$\Delta t_1 = t_1 - t_0$	V_1	I_1	$P_1 = (V_1\, I_1)/1000$	$E_1 = \Delta t_1\, P_1/3600$
t_2	$\Delta t_2 = t_2 - t_1$	V_2	I_3	$P_2 = (V_2\, I_2)/1000$	$E_2 = \Delta t_2\, P_2/3600$
t_3	$\Delta t_3 = t_3 - t_2$	V_3	I_3	$P_3 = (V_3\, I_3)/1000$	$E_3 = \Delta t_3\, P_3/3600$
t_n	$\Delta t_n = t_n - t_{n-1}$	V_n	I_n	$P_n = (V_n\, I_n)/1000$	$E_n = \Delta t_n\, P_n/3600$
					$E = E_1 + E_2 + E_3 + \cdots + E_n$

4. Incubate plate upright at 25 °C for 120–168 h.
5. Count the plate with 10–150 colonies.
6. Perform each test in duplicate. Express the results as CFU/mL.

2.4.6 Heat Generation and Energy Consumption Calculation

1. Electric power (P) was expressed as current (I) multiplied by voltage (V). Energy consumption (E) was calculated as the integral of the product of V and I over time (t) [15]:

$$P = IV \tag{3}$$

$$E = \int_0^t P(t)\mathrm{d}t = \int_0^t VI(t)\mathrm{d}t \cong \sum_0^t \Delta t VI(t) \tag{4}$$

2. Fill obtained data in the following Table 3.

2.4.7 Comparison on Product Properties, Microbial Counts, and Energy Consumption

Fill obtained data in Table 4.

2.5 Determination of Vitamin C Degradation in Acid Juice [15]

In addition to food safety, nutrient loss is of great concern. Ohmic heating has shown less [10, 16, 17], equal [11, 12, 16], and more thermal damage [12] to heat-sensitive compounds. For example, the degradation of vitamin C under ohmic heating was similar to and higher than the conventional heating at low and high electric field strength, respectively [12]. Here we describe the step to investigate vitamin C degradation using static ohmic heating and conventional heating.

1. Mix orange juice with ascorbic acid at the ratio of 20 mg of ascorbic acid and 100 g of juice. Take the sample and use it as an untreated sample.
2. Fill the mixture to the ohmic chamber and turn on the static ohmic heating system (*see* **Note 16**).

Table 4
A possible template for organizing data related to properties, microbial counts, and energy consumption in an ohmic study

		Untreated	Conventional heating	Ohmic heating
Product properties	pH			
	Total soluble solid content			
	ΔE	–		
Microbial counts	Total plate counts (CFU/mL)			
	Yeast and mold counts (CFU/mL)			
Energy consumption (kWh)				

3. After treatment, add 3% (w/v) of meta-phosphoric acid solution and sonicate for 10 min using an ultrasonic bath.
4. Fill the mixture with 3% (w/v) of meta-phosphoric acid solution to 50 ml and filter through 0.45 μm syringe membrane filter.
5. Inject 20 μm into the high-performance liquid chromatography using a reverse C_{18} Hypersil ODS (5 μm, 250 × 4.0 mm) at 25 °C. Use an isocratic gradient of 0.3 mM KH_2PO_4 in 0.35% (v/v) of H_3PO_4 as the mobile phase with the flow rate of 0.8 ml/min for 7 min in which diode array detection is carried on at 248 nm.
6. Heat the orange juice using the conventional heating technique and repeat the step.
7. Compare ascorbic acid content before and after ohmic heating and conventional heating.

3 Notes

1. Establish safety standards for each equipment. All equipment should be mounted on a sturdy flat surface. Before working, ensure that all equipment and wires are correctly connected and grounded and high-voltage areas are covered.
2. Work safety at all times. Pay careful attention to the danger of electricity; for example, do not contact high-voltage areas. Wear electrical insulating gloves for electrical protection during operation.

3. An ohmic heating system is mainly composed of the power supply system, an ohmic heating cell/chamber, electrodes, measuring devices (temperature, current, and voltage), and data acquisition connecting to the personal computer.
4. Static ohmic heating cell/chamber can be constructed into many geometries such as T-cell shape, vertical cylinder, and rectangle. Different non-conductive materials can be used, such as glass, polypropylene, polyethylene, Ultem, Ertacetal, and Teflon. In case of sterilization, the lid should be installed and is equipped with ports at least for thermocouples, an air inlet to build up pressure, and a safety valve. Ports for two plate electrodes, manual valves, and a pressure gauge can be provided, which depends on the design.
5. Electrodes can be designed in different shapes that cooperate with the ohmic heating cell's geometries and can be made of stainless steel, titanium, and titanium coated with platinum. Also, in continuous mode of process, the number of electrodes could be more than 2. Also, the type of electrodes affects the electrochemical reactions and corrosion rate.
6. Power supply system generally consists of a variable transformer, an isolation transformer [18], and electrical safety devices such as circuit breakers and a safety switch.
7. Fiber optic is another option.
8. Accuracy of temperature measuring devices is essential. Thermocouples should be coated with Teflon or ceramic to avoid electrical interference. Signal isolation is also recommended [18].
9. Calibration should be conducted for every measurement device before operation.
10. For the measurement of electrical conductivity, the T-cell and rectangular chambers are recommended. Length L and cross-sectional area A of the sample are used to calculate the electrical conductivity using Eq. (1). Electrical conductivity should be calibrated using at least three different concentrations of standard solution.
11. Example of ohmic heating cell setup: Inside diameter and length of the T-cell chamber made of Pyrex glass is 0.025 m and 0.14 m. Two electrodes with 0.025 m in diameter are connected at both ends of cylindrical spacers made of non-conductive materials. O-ring should be carefully designed to fit the T-cell chamber and seal the ends of the tube. Set the distance between two electrodes (L) to be 0.025 m. Target electric field ranges from 20 to 30 V/cm
12. Electric field strength (E) is calculated as voltage (V) over the distance between two electrodes (L). The E can be varied by

adjusting applied voltage and L. The unit can be V/cm and V/m.

$$E = \frac{V}{L}$$

13. Ensure the contact between electrodes and sample is achieved.
14. A rectangular chamber and a T-cell chamber are recommended.

    Rectangular chamber: The lid is equipped with ports for thermocouples and two electrodes.

    T-cell chamber: Inside diameter and length are 0.05 m and 0.20 m, respectively. Two electrodes with 0.05 m in diameter are connected at both ends of cylindrical spacers. The distance between the two electrodes is 0.10 m. A rubber stopper at the top opening is for inserting thermocouple.
15. Please check the electrical conductivity of solid particles and salt solution to ensure that particulates are less or more conductive than the medium. Food formulation to adjust electrical conductivity is recommended.
16. An ohmic heating cell having water jacket can be used.
17. Install current transducer which links to data acquisition and personal computer.
18. The pH meter needs to be calibrated with buffer solution with three different pH values before use.
19. The spectrophotometer should be standardized using a black glass and a white porcelain calibrated tile before working.
20. Adequate training in standard and specialized microbiological techniques. This is no substitute for proper training in microbiology lab techniques.
21. Place all the media and utensils in an autoclave and sterilize them at 121 °C for 15 min.
22. Conduct experiment under strict aseptic conditions of Biosafety Cabinet.

References

1. de Alwis AAP, Fryer PJ (1992) Operability of the ohmic heating process: electrical conductivity effects. J Food Eng 15(1):21–48. https://doi.org/10.1016/0260-8774(92)90038-8
2. Sastry SK, Kamonpatana P (2014) Electrical conductivity of foods. In: Engineering properties of foods, 4th edn. CRC Press, Boca Raton, pp 527–570. https://doi.org/10.1201/b16897
3. Shynkaryk MV, Ji T, Alvarez VB, Sastry SK (2010) Ohmic heating of peaches in the wide range of frequencies (50 Hz to 1 MHz). J Food Sci 75(7):E493–E500
4. Zell M, Lyng JG, Cronin DA, Morgan DJ (2009) Ohmic heating of meats: electrical conductivities of whole meats and processed meat ingredients. Meat Sci 83(3):563–570
5. Kamonpatana P (2012) Mathematical modeling and microbiological verification of ohmic heating of solid-liquid mixture in continuous

flow Ohmic heater systems. The Ohio State University, Columbus

6. Sarang S, Sastry SK, Gaines J, Yang TCS, Dunne P (2007) Product formulation for ohmic heating: blanching as a pretreatment method to improve uniformity in heating of solid-liquid food mixtures. J Food Sci 72(5): E227–E234. https://doi.org/10.1111/j.1750-3841.2007.00380.x
7. Gavahian M, Farahnaky A (2018) Ohmic-assisted hydrodistillation technology: a review. Trends Food Sci Technol 72:153–161. https://doi.org/10.1016/j.tifs.2017.12.014
8. Mercali GD, Gurak PD, Schmitz F, Marczak LDF (2015) Evaluation of non-thermal effects of electricity on anthocyanin degradation during ohmic heating of Jaboticaba (Myrciaria cauliflora) juice. Food Chem 171:200–205. https://doi.org/10.1016/j.foodchem.2014.09.006
9. Tiravibulsin C, Lorjaroenphon Y, Udompijitkul P, Kamonpatana P (2021) Sterilization of coconut milk in flexible packages via ohmic-assisted thermal sterilizer. LWT 147: 111552. https://doi.org/10.1016/j.lwt.2021.111552
10. Vikram VB, Ramesh MN, Prapulla SG (2005) Thermal degradation kinetics of nutrients in orange juice heated by electromagnetic and conventional methods. J Food Eng 69(1): 31–40. https://doi.org/10.1016/j.jfoodeng.2004.07.013
11. Jaeschke DP, Marczak LDF, Mercali GD (2016) Evaluation of non-thermal effects of electricity on ascorbic acid and carotenoid degradation in acerola pulp during ohmic heating. Food Chem 199:128–134. https://doi.org/10.1016/j.foodchem.2015.11.117
12. Mercali GD, Jaeschke DP, Tessaro IC, Marczak LDF (2012) Study of vitamin C degradation in acerola pulp during ohmic and conventional heat treatment. LWT Food Sci Technol 47(1):91–95
13. Maturin L, Peeler JT (2001) Bacteriological Analytical Manual (BAM) Chapter 3: aerobic plate count. US Food and Drug Administration. Retrieved from https://www.fda.gov/food/laboratory-methods-food/bam-chapter-3-aerobic-plate-count
14. Tournas V, Stack ME, Mislivec PB, Koch HA, Bandler R (2001) BAM chapter 18: yeasts, molds and mycotoxins
15. Wattanayon W, Udompijitkul P, Kamonpatana P (2021) Ohmic heating of a solid-liquid food mixture in an electrically conductive package. J Food Eng 289:110180. https://doi.org/10.1016/j.jfoodeng.2020.110180
16. Assiry A, Sastry SK, Samaranayake C (2003) Degradation kinetics of ascorbic acid during ohmic heating with stainless steel electrodes. J Appl Electrochem 33(2):187–196. https://doi.org/10.1023/A:1024076721332
17. Lee SY, Sagong HG, Ryu S, Kang DH (2012) Effect of continuous ohmic heating to inactivate Escherichia coli O157: H7, Salmonella typhimurium and Listeria monocytogenes in orange juice and tomato juice. J Appl Microbiol 112(4):723–731. https://doi.org/10.1111/j.1365-2672.2012.05247.x
18. Sarang SS, Heskitt BF, Sastry SK (2014) Ohmic heating laboratory units. Ohmic heating in food processing, pp 177–182. https://doi.org/10.1201/b16605

Chapter 13

Microwave Processing: Methods and Procedures Related to Process Parameters

Amandine Rousset, Emmanuel Petitcolas, Farid Chemat, and Sandrine Perino

Abstract

Food and natural products are complex combinations of metabolites and other organic and mineral compounds. Different methods can be used to extract these metabolites: Soxhlet, maceration, distillation, etc. Nevertheless, all these extraction methods consume solvent such as hexane or mixtures of hydrocarbons. Over the last 10 years, there has been a growing demand for new extraction techniques with shortened extraction times and reduced organic solvent consumption, in order to limit contamination and cut costs. This chapter provides an overview of microwave-assisted and solvent-free extraction methods. Microwave energy offers a solution in achieving the objective of sustainable and green chemistry for research, teaching, and commercial applications. Solvent-free microwave extraction (SFME) and microwave hydrodiffusion and gravity (MHG) are described for the extraction of orange essential oil and for the extraction of polyphenols from grape berry by-products.

Key words Microwave, Extraction, Green process, Essential oil, Secondary metabolites

1 Introduction

Microwaves are electromagnetic waves created by magnetrons. They have a wide frequency interval, ranging from 300 MHz to 300 GHz. These waves are a source of energy and cause molecules to move, especially water. When the water molecules are under the influence of microwaves, they rotate, and this stirring produces heat. In general, frequencies around 2.45 GHz are the ones used worldwide for industrial and scientific purposes. In contrast with conventional heating, microwave heat transfer is not limited to thermal conduction or convection currents. In practice, this means that a much faster temperature increase can be obtained. This special feature offers new possibilities in the food industry and more specifically in the extraction field. Processes can be done rapidly, without any added solvent and without wastewater

Mohsen Gavahian (ed.), *Emerging Food Processing Technologies*,
Methods and Protocols in Food Science, https://doi.org/10.1007/978-1-0716-2136-3_13,

production [1]. The advantages of using microwave energy, relative to conventional processing, is that it provides more effective heating, faster startup, faster energy transfer, reduced thermal gradients, and better heating control and eliminates processing steps, which results in increased production [2]. In this chapter, two applications in food thermal processing will be described, both without added solvent: solvent-free microwave extraction (SFME) of orange essential oil and microwave hydrodiffusion and gravity (MHG) producing a high polyphenolic content grape berry extract.

Solvent-free microwave extraction (SFME) *is a combination of microwave heating and distillation at atmospheric pressure, using no added water [3]. The essential oil contained in the plant is evaporated by the in situ water of the plant material [4]. A cooling system outside the microwave oven condenses the distillate continuously. The water excess is returned to the extraction vessel in order to restore the in situ water to the plant material.*

Microwave hydrodiffusion and gravity (MHG) *is a green microwave extraction technique, patented by Chemat et al. [5]. It is a combination of microwave heating, dry distillation at atmospheric pressure, and the force of gravity. It involves placing plant material in a microwave reactor, without adding any solvent or water. The microwaves increase the pressure inside the cells by heating the in situ water. The water and molecules contained in the cells diffuse outside the plant material, and gravity causes them to drop out of the microwave reactor [6]. A cooling system outside the microwave oven cools the extract continuously.*

2 Materials

2.1 Solvent-Free Microwave Hydrodistillation (SFME) of Orange Essential Oil

2.1.1 Microwave Extraction

1. Distilled water.
2. 300 g of orange peels.
3. An Erlenmeyer flask.
4. One kitchen blender.
5. A Milestone ETHOS microwave laboratory oven (Italy) with a frequency of 2.45 GHz and variable power (10 W increments, maximal value: 900 W).
6. A microwave reactor and optic fiber sensors. The microwave has to be modified to add a cooler and Clevenger-type apparatus above.

2.1.2 Analysis

A Hewlett Packard 6890 Gas Chromatography system with a fused silica capillary column with an apolar stationary phase HP5MSTM (30 m × 0.25 mm × 0.25 μm film thickness) and a Hewlett Packard 5973A Mass Spectrometer.

2.2 Microwave Hydrodiffusion and Gravity High Polyphenolic Content Grape Berry Extraction

2.2.1 Extraction

1. 1.5 kg of fresh grape berry (including berries and stems).
2. One vertical laboratory hydraulic press (REUS, $P = 200$ bars, France).
3. A Milestone ETHOS microwave laboratory oven (Italy) with a frequency of 2.45 GHz and variable power (10 W increments, maximal value: 900 W).
4. A microwave reactor and optic fiber sensors. The microwave has to be modified to add a cooler below.

2.2.2 Analysis

1. Folin reagent (Sigma-Aldrich) solution: in 100 mL volumetric flask, put 10 mL of Folin reagent and fill up with distilled water to the volumetric line.
2. Na_2CO_3 (Sigma-Aldrich) solution: in a 100 mL volumetric flask, place 7.5 g of Na_2CO_3 and fill up with distilled water to the volumetric line.
3. 10% formic acid (Sigma-Aldrich) solution: in 100 mL volumetric flask, put 10 mL of formic acid and fill up with distilled water to the volumetric line.
4. Malvidin-3-*O*-glucoside (standard quality, Extrasynthese) stock solution: in 50 mL volumetric flask, place 5 mg of malvidin-3-*O*-glucoside and fill up with 10% formic acid solution to the volumetric line.
5. Gallic acid (standard quality, Sigma-Aldrich) stock solution: in a 50 mL volumetric flask, dispense 50 mg of gallic acid and fill up with distillated water to the volumetric line.
6. Distilled water.
7. A Hewlett Packard diode array 8453 UV spectrophotometer with cuvettes.

3 Methods

3.1 Solvent-Free Microwave Hydrodistillation (SFME) of Orange Essential Oil

3.1.1 Ground Orange Peels (Figs. 1, 2, 3, 4, 5, 6, 7, and 8)

Take the orange peels and grind them to obtain pieces between 0.2 and 1 cm (*see* **Note 1**).

3.1.2 Solvent-Free Microwave Extraction

Put the ground orange peels inside the microwave reactor and place the reactor inside the microwave. Put distilled water inside the Clevenger apparatus (to reach the equilibrium) and turn on the cooler (5 °C).

Fig. 1 Orange peels before and after grinding

Fig. 2 Schematic diagram of SFME apparatus

Apply a power of 300 W for 20 min. Pay attention, because as the essential oil is extracted and recovered in the Clevenger apparatus, the same amount of water has to go into the container in order to prevent the appearance of hot spots (*see* **Note 2**). Once the 20 min extraction is completed, you can collect the water in a beaker and the essential oil separately, in a pre-weighed flask. Be careful, as the microwave reactor inside the microwave can still be hot (*see* **Note 3**). Yield can be expressed in percentage of essential oil in the orange peels.

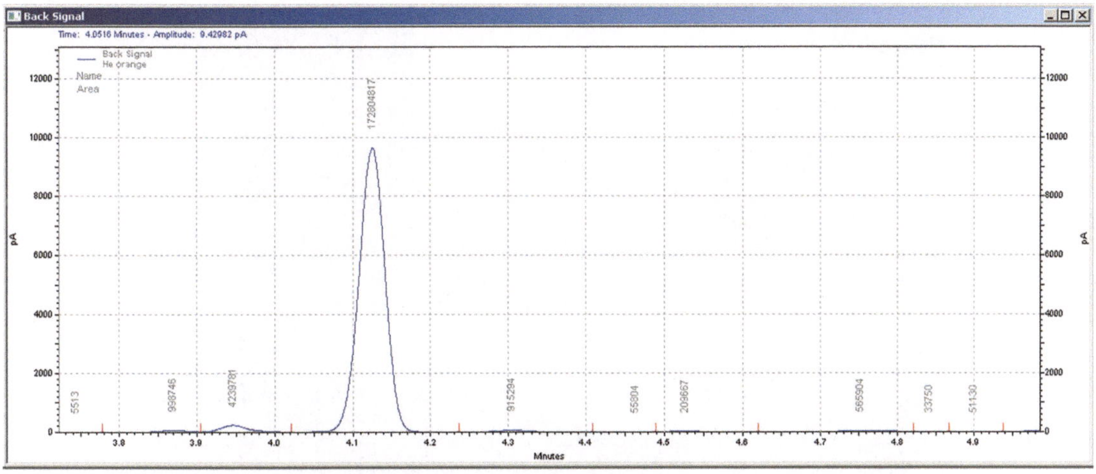

Fig. 3 Orange essential oil chromatograph

Fig. 4 Grapes before and after the hydraulic press

3.1.3 Chromatographic Analysis of the Orange Essential Oil

Gradient to enter in the software associated with the GC: from 60 to 280 °C at 2 °C/min with a flow rate of 0.3 mL/min carrier gas (N_2). Injection has to be performed at 250 °C in the spitless mode and you have to inject 1 μL of sample.

Components of the essential oil can be identified via their relative retention indices on HP-5MS TM column, determined with reference to a homologous series of C5–C28 *n*-alkanes, and by a comparison of their mass spectral fragmentation patterns with those stored in the data bank (Wiley/NBS library) and the literature (*see* **Note 4**).

Fig. 5 Hydraulic press REUS

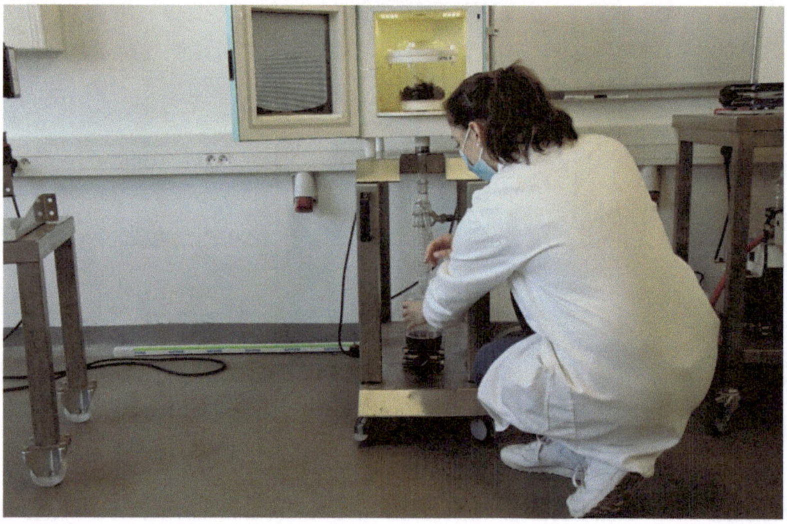

Fig. 6 MHG extraction

3.2 Microwave Hydrodiffusion and Gravity High Polyphenolic Content Grape Berry Extraction

3.2.1 Obtaining the Grape Berry By-Product

Press all the grapes with the hydraulic press. You will obtain two products: on one side the juice (around 600 mL) and on the other the grape by-products (press cake), a by-product of the juice process (around 800 g). This mimics industrial grape processing and the by-product usually obtained.

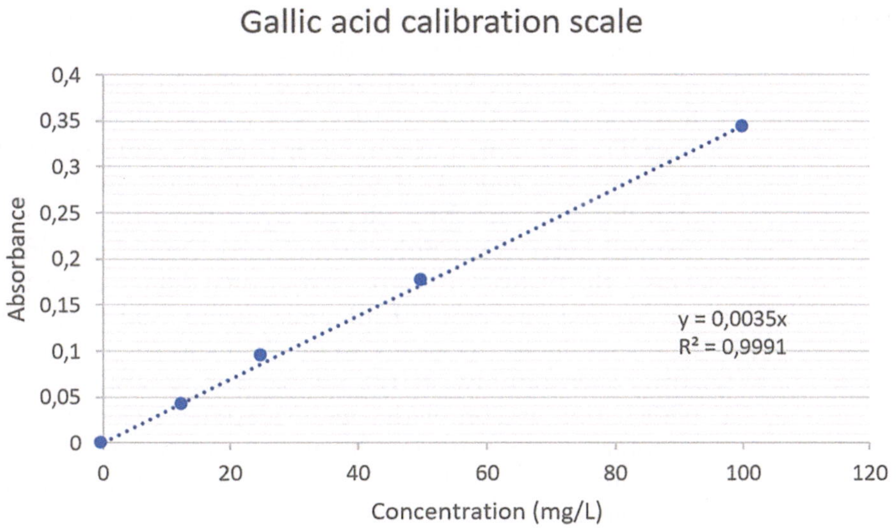

Fig. 7 Gallic acid calibration scale

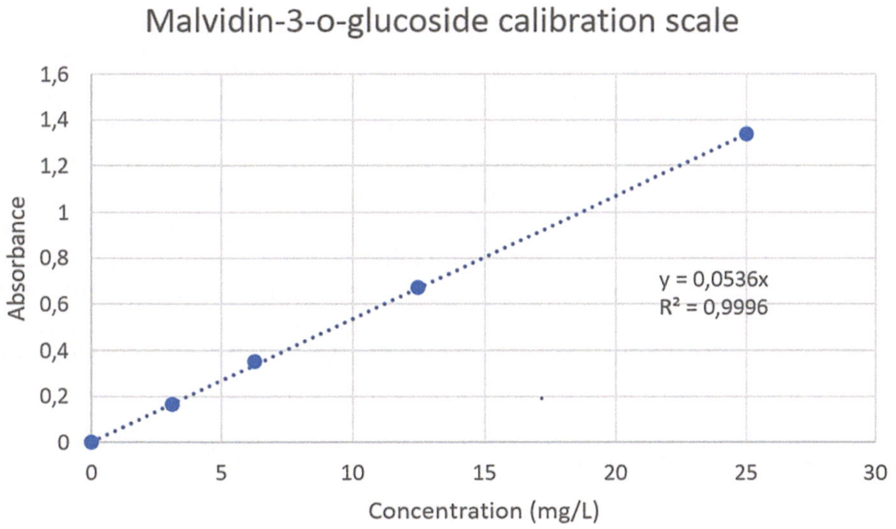

Fig. 8 Malvidin-3-*O*-glucoside calibration scale

3.2.2 Microwave Hydrodiffusion and Gravity Extraction

Put 400 g of grape by-products into the microwave reactor and place an Erlenmeyer flask underneath the cooler. Turn on the cooler (5 °C). Apply a power of 400 W for the duration of 20 min. The extract will flow into the beaker. The grape by-products inside the microwave will still be hot.

3.2.3 Analysis of the Polyphenolic and Anthocyanin Content of the Extract

To determine the percentage of polyphenols and anthocyanins, calibration scales have to be made with standards: gallic acid for polyphenols and malvidin-3-*O*-glucoside for anthocyanins.

Determination of Total Phenolic Content

Dilute the extract obtained by ten by dispensing 1 mL into a 10 mL volumetric flask and adding distilled water up to the line. Take six UV cuvettes (five for calibration and one for your sample). In the first one, put 0.4 mL of gallic acid stock solution and 1.6 mL of Na_2CO_3 solution. Put only 1 mL of Na_2CO_3 in all the other cuvettes. Mix and take 1 mL out of the first cuvette and place it into the next one. Mix. Repeat again three times. Put 1 mL of distilled water in the fifth one and throw away 1 mL after mixing. This is the blank. Put 1 mL of your own diluted extract in the last cuvette and throw away 1 mL after mixing. Add 1 mL of the Folin reagent solution to each cuvette. You should have four cuvettes with concentrations of 100, 50, 25, and 12.5 mg/L, one cuvette with a blank, and one with your own extract. Leave them in the dark at room temperature for 30 min. Absorbances are measured at 760 nm (*see* **Note 5**). Results are expressed as g/L gallic acid equivalent.

Determination of Total Anthocyanin Contents

Following the same procedure as that for the phenolic content, samples must be prepared for the calibration scale. Attention: in this procedure, dilutions have to be made with the 10% formic acid solution. Dilute your extract by 10. Take six UV cuvettes (five for calibration and one for your sample). In the first one, put 0.5 mL of malvidin-3-*O*-glucoside stock solution and 1.5 mL of the 10% formic acid solution. Dispense only 1 mL of the 10% formic acid solution into all the other cuvettes. Mix and take 1 mL out of the first cuvette and place it into the next one. Mix. Repeat again three times. Put 1 mL of distilled water in the fifth one and throw away 1 mL after mixing. This is the blank. Dispense 1 mL of your own diluted extract in the last cuvette. You should have four cuvettes with concentrations of 25, 12.5, 6.25, and 3.125 mg/L, one cuvette with a blank, and one with your own extract. Absorbances are measured at 530 nm (*see* **Note 6**). Results are expressed as mg/mL malvidin-3-*O*-glucoside equivalent.

4 Notes

1. If you grind into very tiny pieces, you can lose some essential oil in the blender. If it is not ground enough, the extraction will be more difficult, and some essential oil will stay inside the peel.
2. Note that if working with a different mass, it is recommended to apply 1 W/g of product as a general rule.
3. With the orange samples used here, we usually found between 0.5% and 2% of essential oil in the orange peels.
4. With the orange samples used here, the essential oil is usually mainly composed of monoterpenes (97%) of which 95% is limonene.

5. The sample will become blue/gray. Verify that the sample is diluted enough: the intensity of the color has to be near to the sample of the calibration scale. Absorbances depend on the Folin solution used: the calibration scale probably will not be exactly the same. For the extract described here, we found 1.4 g/L gallic acid equivalent. In the final calculation, remember that the extract has been diluted by 10.
6. Verify that the sample is diluted enough: the intensity of the color has to be similar to the calibration scale sample. For the extract used here, we found 0.33 g/L malvidin-3-*O*-glucoside equivalent. In the final calculation, remember that the extract has been diluted by 10.

5 Conclusion

Microwave extraction is a green extraction method that offers important advantages like short extraction time (20 min), low energy input, and no requirements in solvents. This chapter proved the efficiency of SFME for the extraction of essential oil and MHG in the extraction of polyphenols and anthocyanins from grape by-products.

Acknowledgments

We thank master's students from Avignon University for their assistance in this green extraction experiment in teaching laboratories.

References

1. Perino-Issartier S, Maingonnat JF, Chemat F (2011) Microwave in food processing. "Green food science and technology, alternatives to conventional food processing". RCS green chemistry book series. RCS, London
2. Perino S, Petitcolas E, Chemat F (2017) Chap 16. Microwave extraction of natural products in the teaching laboratory: fundamentals of essential oils green extraction. In: Microwave chemistry. De Gruyter, Berlin
3. Chemat F, Lucchesi ME, Smadja J (2004) Solvent free microwave extraction of volatile natural substances. US Patent 0187340A1
4. Ferhat MA, Meklati BY, Smadja J, Chemat F (2006) An improved microwave Clevenger apparatus for distillation of essential oils from orange peel. J Chromatogr A 1112:121–126
5. Chemat F, Abert-Vian M, Visinoni F (2008) Microwave hydro-diffusion for isolation of natural products. European Patent EP1955749
6. Perino-Issartier S, Visinoni F, Chemat F (2011) Green chemical processing in the teaching laboratory: microwave extraction of natural products. In: Sanghi R, Singh V (eds) Green chemistry for environmental remediation. Wiley, Chichester

Chapter 14

Infrared Heating Processing: Methods and Procedures Related to Process Parameters

Asaad Rehman Al-Hilphy, Mohsen Gavahian, and Sriram Vidyarthi

Abstract

Infrared (IR) is an electromagnetic wave which lies between the visible light and microwave, and its wavelength ranges between 0.76 and 1000 μm. IR heating is an emerging heating technology which is considered as an eco-friendly technology. This chapter focused on using IR for drying food and investigating the effect of IR on the drying parameters. In addition, the chapter describes methods of calculation of drying characteristics and mass transfer during drying, as well as mass transfer coefficient. To exemplify the drying methods, procedures, and characteristics, strawberry drying has been taken as an example.

Key words Infrared heating, Emerging technologies, Protocols, Drying rate, Mass transfer

1 Introduction

Heat is transferred through conduction, convection, and radiation [1]. The main goals of heating food are to increase shelf life and improve taste [2]. Conventional heating causes physical and chemical changes to the final food product. In conventional heating, the source of heating is the combustion of fuel or electrical heaters. The heat transfer process from the source to the food depends on the type of cooking. For example, in the case of the baking, frying, and boiling processes, the heat energy is transmitted from heating medium to the food surface through convection, whereas inside the food, it is primarily transferred through conduction. The rise in food temperature and the time required for this depend on the thermal and engineering properties of the food [3, 4].

Infrared (IR) is a part of the electromagnetic spectrum that lies between the visible light and microwave, and its wavelength ranges between 0.76 and 1000 μm [5–7]. Based on different applications, IR band has been mainly divided into three subsections. One of the most commonly applied IR classifications is near-IR (NIR), mid-IR (MIR), and far-IR (FIR) radiations with wavelength ranges of

Mohsen Gavahian (ed.), *Emerging Food Processing Technologies*,
Methods and Protocols in Food Science, https://doi.org/10.1007/978-1-0716-2136-3_14,

Table 1
IR radiation wavelengths and corresponding black body peak temperatures [12]

IR division	Wavelength (μm)	Temperatures (°C)
Near infrared (NIR) or short wave (SW)	0.76–2	3350–1175
Medium infrared (MIR) or medium wave (MW)	2–4	1175–450
Far infrared (FIR) or long wave (LW)	4–1000	450–(−270)

0.76–2 μm (3350–1175 °C), 2–4 μm (1175–450 °C), and 4–1000 μm (450–(−270)°C), respectively, as shown in Table 1.

IR heating is an emerging technology that can have applications in several food processing operations, such as drying, cooking, roasting, baking, peeling, blanching, pasteurization, and sterilization [7–12]. Infrared technology is considered environmentally friendly and less-energy and water-intensive process compared to traditional heating. IR dryer generates six to ten times higher heat flux than a conventional convection hot air dryer [14], which can remarkably reduce the drying time. The major advantages of IR heating are rapid heating rate, higher heat transfer rate, higher energy efficiency, and better product quality [14, 15].

IR heating is affected by several key parameters which must be reported to optimize the heating process. These parameters are mainly emitter spectral characteristics and characteristics of materials to be heated.

1.1 Emitter Spectral Characteristics

The spectral characteristics (distribution of radiant energy over the wavelength spectrum) of an emitter are crucial factors that affect the IR heating process significantly. These characteristics are linked to emitter surface temperature, emitter emissivity, emissive power, radiant intensity, peak wavelength, and radiant efficiency. Ideally, the emissive characteristics of the IR emitter and the absorption characteristics of the object to be heated should match in order to achieve optimal heating efficiency. The spectral characteristics of IR emitters are highly dependent on their construction materials, morphology, and operating temperatures.

The emitter temperatures applicable for food processes normally range from 260 to 2200 °C [17]. The thermal effect of radiation is associated with the wavelengths emitted from the heating source [5]. The wavelength at which the maximum radiation intensity occurs is called the peak wavelength, which is determined by the temperature and emissivity of emitter [6]. The relationship between the emitter temperature, emissive power, and peak wavelength is described by Planck's distribution law, Stefan-Boltzmann's law, and Wien's displacement law for blackbody radiation [5, 18]. In general, catalytic IR emitters attain lower power density (usually <22 kW/m^2) due to the lower surface temperature (usually <700 °C) than the electric IR emitters which usually have higher surface temperature and power density [11, 19].

1.2 Characteristics of Material to Be Heated and Heating Environment

The object to be heated and its heating environment during IR heating process are important factors that affect the heating process. Regardless of IR radiation intensity of the emitter, the final radiation intensity reaching the surface of the object plays a crucial role in IR heating performance. In addition, the thermophysical properties of the food product greatly affect the IR processing performance and processed product quality. Materials absorb IR energy through a mechanism of molecular vibration and rotation. The food material IR absorptivity and emissivity are associated with its physicochemical nature, water content, and thickness [20]. The emissivity of water and most organic components of food ranges from 0.78 to 0.95 [21]. When IR radiation hits on the food surface, its constituent organic materials absorb the radiation at distinct frequencies based on intramolecular changes in vibrational state (2.5–100 μm, MIR to FIR) and rotational state (>100 μm, FIR) [3, 6]. Usually, food materials absorb MIR and FIR more efficiently through stretching vibration, which generates heat energy. After absorption of IR radiation, food materials emit thermal radiation based on their emissivity, which affects the overall heating process of food.

As an example, in order to elaborate the IR process, the next part of this chapter will briefly present the parameters of strawberry drying using infrared heating based on our previous study [22]. For detailed information on materials and procedures, the above study is greatly recommended. The procedures and methods can be applied to any other fruits and vegetables with required modifications.

2 Materials

1. 1500 g of strawberries.
2. Digital infrared halogen dryer (halogen lamp emits infrared radiation) model EL-817D, Mgf. Co., Ltd., China (*see* **Note 1**).
3. Solution of 2% sodium metabisulfite (*see* **Note 2**).
4. Oven (LTE Scientific, United Kingdom) for determination of moisture content of fresh strawberry at a temperature of 105 °C.
5. Sensitive balance (Kern/Denver, Germany).

3 Methods

3.1 Determination of Moisture Content during IR Drying of Strawberries

1. Cut 500 g of strawberry radially to slices with 1 cm thickness.
2. Immerse slices into the sodium metabisulfite solution for 10 min.
3. Spread slices on the tray located inside the halogen oven.
4. Turn on the halogen oven and set up a temperature of 60 °C.
5. Record drying time at an interval 5 min of any appropriate time interval and weight of the product using a sensitive balance. Record the data in a tabular form till an equilibrium weight is achieved (i.e., no further change in weight), as shown in Table 2.
6. Calculate initial moisture content (M_i) of strawberries according to the AOAC Official Methods of Analysis [23].
7. Calculate the dry matter of strawberries from the following equations:

$$W_s\,(\%) = 100\% - M_i(\%) \tag{1}$$

$$W_s\,(g) = W_0 * W_s\,(\%) \tag{2}$$

8. Calculate moisture content (%) with drying time as shown in Table 3.
9. Calculate moisture content in dry basis (d.b.) as shown in Eq. (3) and Table 2.

$$M\left(\frac{\text{kg water}}{\text{kg d.b.}}\right) = \frac{M\,(\%)}{100 - M(\%)} \tag{3}$$

10. Repeat the steps from (i) to (vi) for temperatures 70 and 80 °C.
11. Establish the relationship between drying time (min) and moisture content $\left(\frac{\text{kg water}}{\text{kg d.b.}}\right)$ in excel program as illustrated in Fig. 1.

Table 2
Changing weight of product during drying with drying time

Drying time (min)	Weight of product (g)
t_0	W_{t0}
t_1	W_{t1}
t_2	W_{t2}
t_3	W_{t3}
.	.
.	.
.	.
t_n	W_{tn}

Table 3
Calculation of moisture content in wet basis (%) and in dry basis $\left(\frac{\text{kg water}}{\text{kg d.b.}}\right)$ and drying time

Drying time (min)	Moisture content (%)	Moisture content $\left(\frac{\text{kg water}}{\text{kg d.b.}}\right)$
t_0	$M_{t0} = \frac{W_{t0}-W_s}{W_{t0}} \times 100$	$M_{t0} = \frac{M_{t0}\ (\%)}{100-M_{t0}\ (\%)}$
t_1	$M_{t1} = \frac{W_{t1}-W_s}{W_{t1}} \times 100$	$M_{t1} = \frac{M_{t1}\ (\%)}{100-M_{t}\ (\%)}$
t_2	$M_{t2} = \frac{W_{t2}-W_s}{W_{t2}} \times 100$	$M_{t2} = \frac{M_{t2}\ (\%)}{100-M_{t2}\ (\%)}$
t_3	$M_{t3} = \frac{W_{t3}-W_s}{W_{t3}} \times 100$	$M_{t3} = \frac{M_{t3}\ (\%)}{100-M_{t3}\ (\%)}$
. . .	. . .	. . .
t_n	$M_{tn} = \frac{W_{tn}-W_s}{W_{tn}} \times 100$	$M_{tn} = \frac{M_{tn}\ (\%)}{100-M_{tn}\ (\%)}$

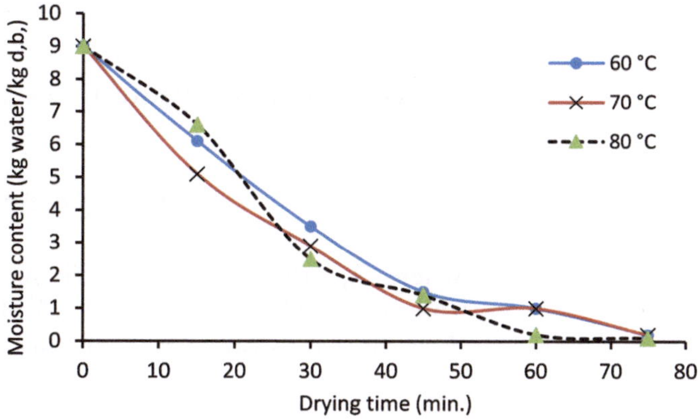

Fig. 1 Drying curves of strawberry with halogen dryer at different temperatures

3.2 Determination of Effective Moisture Diffusion and Activation Energy

1. The moisture ratio can be calculated using diffusion equation following Crank-Nicolson method [24] as shown in Eq. (4) (*see* **Note 3**):

$$\text{MR} = \frac{M - M_e}{M_o - M_e} = \frac{8}{\pi^2} \sum_{n=0}^{\infty} \frac{1}{(2n+1)^2} \exp\left[\frac{-(2n+1)^2 \pi^2 D_{\text{eff}}.t}{4L^2}\right] \tag{4}$$

where MR is the moisture ratio, M is the moisture content (kg water/kg db) at time t (s), M_o is the initial moisture content (kg water/kg db), M_e is the equilibrium moisture content (kg water /kg db), L is the slab thickness of strawberry slices (m), and $D_{\text{eff.}}$ is the effective moisture diffusion (m^2/s).

When drying times are long, the Eq. (4) was simplified to Eq. (5) as follows:

$$\text{MR} = \frac{M - M_e}{M_o - M_e} = \frac{8}{\pi^2} \exp\left[\frac{-\pi^2 D_{\text{eff.}} t}{4L^2}\right] \quad (5)$$

The moisture ratio (MR) is further simplified to M/M_o as the value of M_e is very small compared to M_o and M and can be neglected. Therefore, Eq. (5) can be rewritten as follows:

$$\ln\left(\frac{M}{M_o}\right) = \ln\left(\frac{8}{\pi^2}\right) - \left[\frac{\pi^2 D_{\text{eff.}} t}{4L^2}\right] \quad (6)$$

Diffusion coefficient is calculated by plotting $\ln\left(\frac{M-M_e}{M_o-M_e}\right)$ versus drying time (Fig. 2) (*see* **Note 4**) where slope is shown in Eq. (7):

$$\text{slope} = \frac{\pi^2 D_{\text{eff.}}}{4L^2} \quad (7)$$

The relationship between $D_{\text{eff.}}$ and drying temperature was as below:

$$D_{\text{eff.}} = D_o \exp\left[-\frac{E_a}{RT}\right] \quad (8)$$

where D_o is the constant of Arrhenius equation (m^2/s), R is the gas constant (8.31451 kJ/mol K), T is the drying temperature (K), and E_a is the activation energy (kJ/mol).

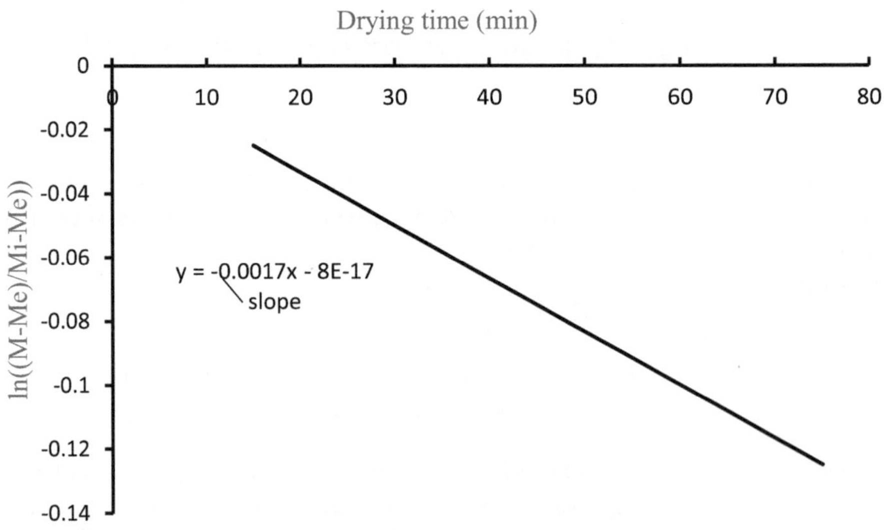

Fig. 2 Moisture content ratio vs. drying time to determine diffusion coefficient ($D_{\text{eff.}}$)

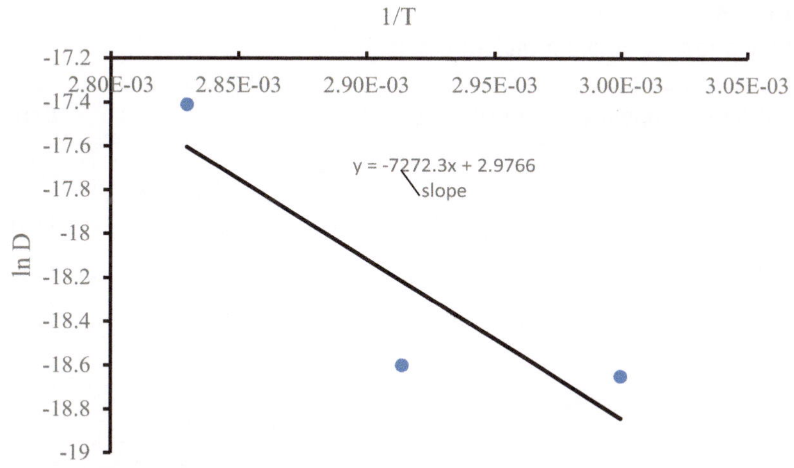

Fig. 3 Relationship between effective moisture diffusivity and temperature using Arrhenius equation

Equation (8) can be rewritten as Eq. (9) as follows:

$$\ln(D_{\text{eff.}}) = \ln(D_{\text{o}}) - \left[\frac{E_{\text{a}}}{RT}\right] \tag{9}$$

where

$$\text{slope} = \frac{E_{\text{a}}}{RT}; \quad E_{\text{a}} = \text{slope} \times RT \tag{10}$$

Activation energy (E_{a}) can be obtained from plotting effective moisture diffusion ($\ln D_{\text{eff.}}$) versus reciprocal drying temperature ($1/T$) (K) where D_{o} value is the intercept (Fig. 3).

3.3 Calculation of Drying Rate

Drying rate is given by the following equation:

$$\text{DR} = \frac{M_{t+\text{d}t} - M_t}{\text{d}t} \tag{11}$$

where DR is the drying rate (kg/min), $M_{t\,+\,\text{d}t}$ is the moisture content (kg water/kg db) at t + dt, t is the drying time (min.), and M_t is the moisture content (kg water/kg d.b.) at time t (min).

Drying rate can also be presented in unit of (kg of water/(m^2-min)) where A is the tray area (m^2) as depicted in Eq. (12) and Table 4:

$$\text{DR} = \frac{M_{t+\text{d}t} - M_t}{\text{d}t \ \ A} \tag{12}$$

Draw the relationship between drying time and drying rate using excel program (Fig. 4).

Table 4
Drying rate at different moisture contents

Moisture content (kg water/kg d.b.)	Drying rate (kg water/(m^2 min))
M_{t0}	$\text{DR}_0 = \frac{M_{t+dt}-M_t}{dt\ A} = 0$
M_{t1}	$\text{DR}_1 = \frac{M_{t0}-M_{t1}}{t_1\ A}$
M_{t2}	$\text{DR}_2 = \frac{M_{t1}-M_{t2}}{t_2\ A}$
M_{t3}	$\text{DR}_3 = \frac{M_{t2}-M_{t3}}{t_3\ A}$
. . .	. . .
M_{t6}	$\text{DR}_6 = \frac{M_{t5}-M_{t6}}{t_6\ A}$
M_{tn}	$\text{DR}_{tn} = \frac{M_{t6}-M_{tn}}{t_n\ A}$

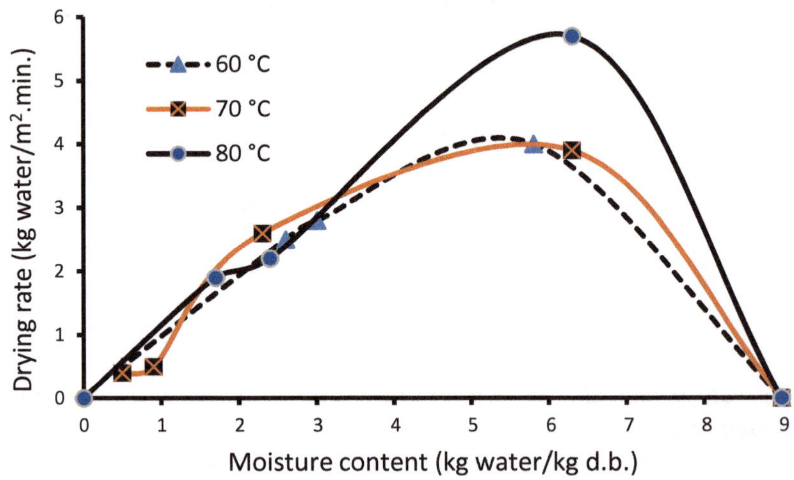

Fig. 4 Variation in drying rate with moisture content of strawberries at different temperatures (*see* **Note 5**)

3.4 Determination of Critical Moisture Content and Drying Rate Periods

The critical moisture content, critical point, constant rate period, falling rate period, and critical time can be determined graphically as illustrated in the Figs. 5 and 6.

3.5 Analytical Calculation of Drying Rates

Drying time during constant rate period can be calculated from the following equation:

$$t_c = \frac{M_{t0} - M_c}{R_c} \tag{13}$$

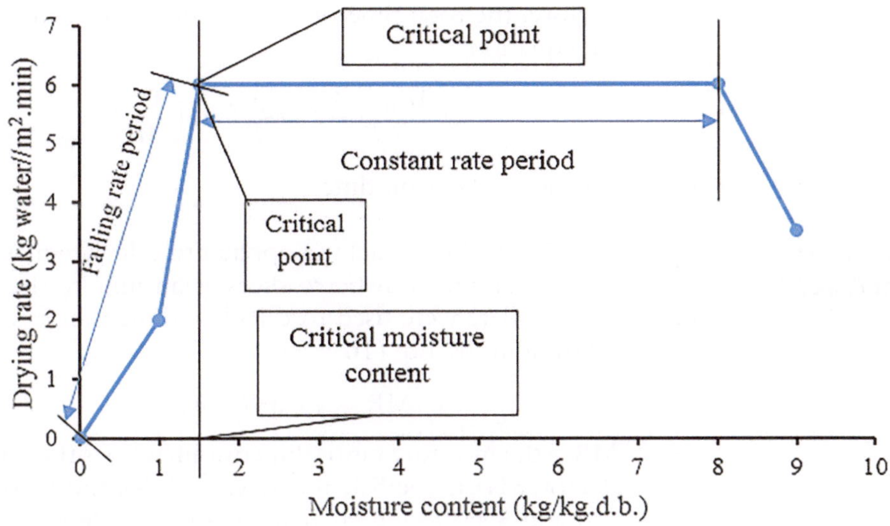

Fig. 5 Determination of critical moisture content and drying rate periods

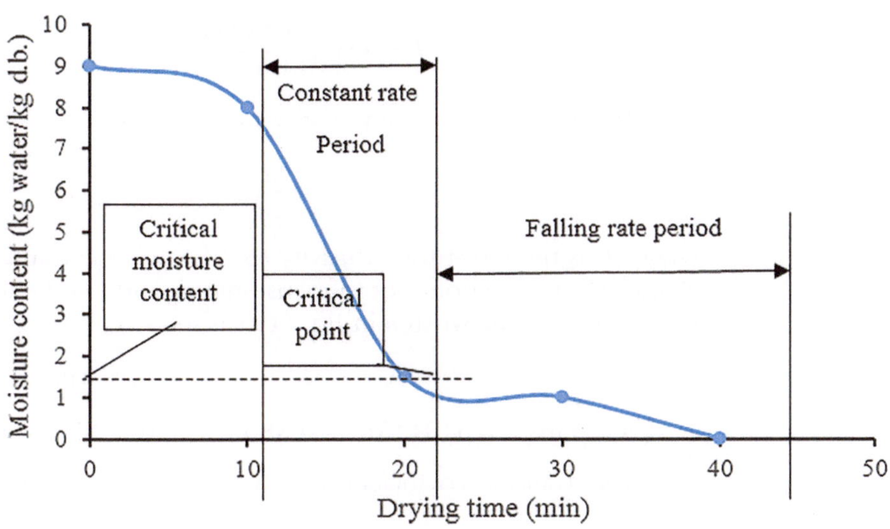

Fig. 6 Determination of critical moisture content by graphical method

where R_c, M_c, and t_c are the drying rate at constant rate period, critical moisture content, and constant rate time, respectively.

Time of falling rate period can be calculated from the following equation:

$$t_f = \frac{M_c}{R_c} \ln\left(\frac{M_c}{M}\right) \tag{14}$$

where R_c, M_c, t_f, and M are the drying rate at falling rate period, critical moisture content, falling rate time, and moisture content after time t in the falling rate period, respectively.

Therefore, the total time of drying can be calculated from the following equation:

$$t = \frac{M_{t0} - M_c}{R_c} + \frac{M_c}{R_c} \ln\left(\frac{M_c}{M}\right) \quad (15)$$

where t is the total drying time.

3.6 Mass Transfer Coefficient Calculation

Dincer and Dost [25] model was applied to calculate mass transfer coefficient for drying strawberry slices as an infinite slab. Experimental drying data were used to calculate dimensionless moisture ratio by fitting in the Eq. (16):

$$\mathrm{MR} = J \exp\,(-St) \quad (16)$$

where MR is the moisture ratio (dimensionless), J is the dimensionless lag factor, S is the coefficient of drying (1/s), and t is the drying time (s). J and S are found via mathematical modeling depending on the experimental moisture ratio and Eq. (16). Biot number (Bi) is given in Eq. (17):

$$J = \exp\left(\frac{0.2533\mathrm{Bi}}{1.3 + \mathrm{Bi}}\right) \quad (17)$$

Moisture diffusivity can be given in Eq. (18):

$$D = \frac{SL^2}{\mu^2} \quad (18)$$

where D is the moisture diffusivity (m^2/s), μ is the transcendental characteristic equation root (dimensionless), and L is the slab thickness (m). μ can be written in Eq. (19) as follows:

$$\begin{cases} \mu = \frac{\pi}{2} & \mathrm{Bi} \geq 100 \\ \mu = \tan^{-1}(0.640443\mathrm{Bi} + 0.380379) & 0 < \mathrm{Bi} < 100 \end{cases} \quad (19)$$

Mass transfer coefficient can be calculated by Eq. (20) as follows:

$$k_m = \frac{\mathrm{Bi}\ D}{L} \quad (20)$$

where k_m is the mass transfer coefficient (m/s).

4 Notes

1. Digital infrared halogen dryer consists of a 12-liter Pyrex glass container, a stainless steel tray for holding the samples, a cover which includes a 1300 W halogen lamp, and a fan for air circulation. The temperature ranges between 0 and 300 °C; input voltage and frequency are 220 V and 50 Hz, respectively.

2. The solution is prepared by weighing 2 g of the substance (sodium metabisulfite) and dissolving it in a fixed amount of distilled water (20 ml) in a small beaker. Then the solution is transferred to a 100 ml bottle, and distilled water is added to a marked volume. This ratio is called a weight/volume ratio.
3. Fick's second law is used to describe the drying behavior of an infinite slab with assumption that diffusion causes the moisture movement. In addition, temperature and diffusion coefficient are constant with negligible shrinkage as suggested by Crank [24].
4. Diffusion coefficient can be calculated for each drying temperature used such as 60, 70, and 80 °C (dry strawberry at different temperatures).
5. In some food, only falling rate period is observed at all drying temperatures. In other food, both constant and falling rate periods are noticed. Moreover, in some food, the falling rate period consists of more than one curve.

References

1. Saravacos GD, Kostaropoulos AE (2002) Handbook of food processing equipment. Springer, New York
2. Krishnamurthy K, Khurana HK, Soojin J, Irudayaraj J, Demirci A (2008) Infrared heating in food processing: an overview. Compr Rev Food Sci Food Saf 7(1):2–13. https://doi.org/10.1111/j.1541-4337.2007.00024.x
3. Rosenthal I (2012) Electromagnetic radiations in food science. Springer-Verlag, Germany.
4. Pan Z, Atungulu GG (2010) Infrared heating for food and agricultural processing. CRC Press, Boca Raton
5. Siegel R (2001) Thermal radiation heat transfer. CRC Press, Boca Raton
6. Sakai N, Hanzawa T (1994) Applications and advances in far-infrared heating in Japan. Trends Food Sci Technol 5(11):357–362
7. Skjöldebrand C (2001) Infrared heating. In: Richardson P (ed) Thermal technologies in food processing. CRC Press, New York
8. Aboud SA, Altemimi AB, Al-HiIphy ARS, Yi-Chen L, Cacciola F (2019) A comprehensive review on infrared heating applications in food processing. Molecules 24(22):4125. https://doi.org/10.3390/molecules24224125
9. Lee E-H (2020) A review on applications of infrared heating for food processing in comparison to other industries
10. Li X (2012) A study of infrared heating technology for tomato peeling: process characterization and modeling. University of California, Davis
11. Li X, Pan Z (2014) Dry peeling of tomato by infrared radiative heating: part II. Model validation and sensitivity analysis. Food Bioprocess Technol 7(7):2005–2013
12. Vidyarthi SK (2017) Study and modeling of infrared heating for tomato dry-peeling. University of California, Davis
13. Vidyarthi S, Li X, Pan Z (2019) Peeling of tomatoes using infrared heating technology. Tomato Chem Ind Process Prod Dev:180–200
14. Therien N, Cote B, Broadbent AD (1991) Statistical analysis of a continuous infrared dryer. Text Res J 61(4):193–202
15. Bekki E (1991) Rough rice drying with a far-infrared panel heater. J Jpn Soc Agric Mach 53(1):55–63
16. Ginzburg AS (1969) Application of infra-red radiation in food processing
17. Sheridan P, Shilton N (1999) Application of far infra-red radiation to cooking of meat products. J Food Eng 41(3–4):203–208
18. Modest MF (2013) Radiative heat transfer. Academic Press, Boston
19. Vidyarthi SK, El Mashad HM, Khir R, Zhang R, Tiwari R, Pan Z (2019) Evaluation of selected electric infrared emitters for tomato peeling. Biosyst Eng 184:90–100

20. Sandu C (1986) Infrared radiative drying in food engineering: a process analysis. Biotechnol Prog 2(3):109–119
21. Fellows PJ (2009) Food processing technology: principles and practice. Elsevier, Amsterdam
22. Al-Hilphy ARS, AlRikabi AKJ (2013) Mathematical modeling and experimental study on thin layer halogen dryer of strawberry and study its effect on antioxidant activity. Am J Agric Biol Sci 8(4):268. https://doi.org/10.3844/ajabssp.2013.268.281
23. AOAC. Official methods of anaylsis. [Online]. http://www.eoma.aoac.org/methods/info.asp?ID=16264
24. Crank J (1979) The mathematics of diffusion. Oxford University Press, Oxford
25. Dincer I, Dost S (1996) A modelling study for moisture diffusivities and moisture transfer coefficients in drying of solid objects. Int J Energy Res 20(6):531–539. https://doi.org/10.1002/(SICI)1099-114X(199606)20:6<531::AID-ER171>3.0.CO;2-6

Chapter 15

Protocol in Food Extrusion Technology

Girish N. Mathad, Jenshinn Lin, and Minh Khoa Nguyen

Abstract

Extrusion technology is mainly used in preparation of various food products such as ready-to-eat breakfast cereals, snack foods, baby foods and texturized vegetable protein (TVP), or meat analogs. This chapter gives overview on extrusion processing, different types of extruders, and its methodology. Part of the chapter focuses on physicochemical changes which occur during the extrusion process and mathematical models. Food extrusion technology is one of the state-of-the-art technologies used in food industry, which can produce many kinds of products. In this chapter, basic information on food extrusion and its protocols are detailed.

Key words Extrusion, Single-screw extruder, Twin-screw extruder, Texturized vegetable protein, Protocols

1 Introduction

Food extrusion technology is one of the important areas of food processing. It is a combination of various unit operations such as mixing, heating, kneading, shearing, and shaping [1]. Food extrusion is a high-temperature short-time processing technique (HTST) [2], which is used in development of different kind of food products [3]. Heat generated during extrusion is due to shearing effect which causes changes in the raw materials [4]. HTST process ensures the safety of final product by maintaining its nutritional value [5] (Fig. 1).

Extrusion is mainly used in production of pet foods, baby foods, and various cereals and protein-based snacks. According to various studies, starches and cereals are ideal for extrusion processing due to its perfect texture, structure, mouthfeel, and other attributes [6]. Extrusion cooking technology is utilized for the production of protein-rich materials prepared from vegetables, i.e., to prepare meat-like structure, which is one of the main principles of extrusion cooking [7]. Literally, the word "extrusion" is derived from Latin word "extrudere" stating the process of pushing

Mohsen Gavahian (ed.), *Emerging Food Processing Technologies*,
Methods and Protocols in Food Science, https://doi.org/10.1007/978-1-0716-2136-3_15,

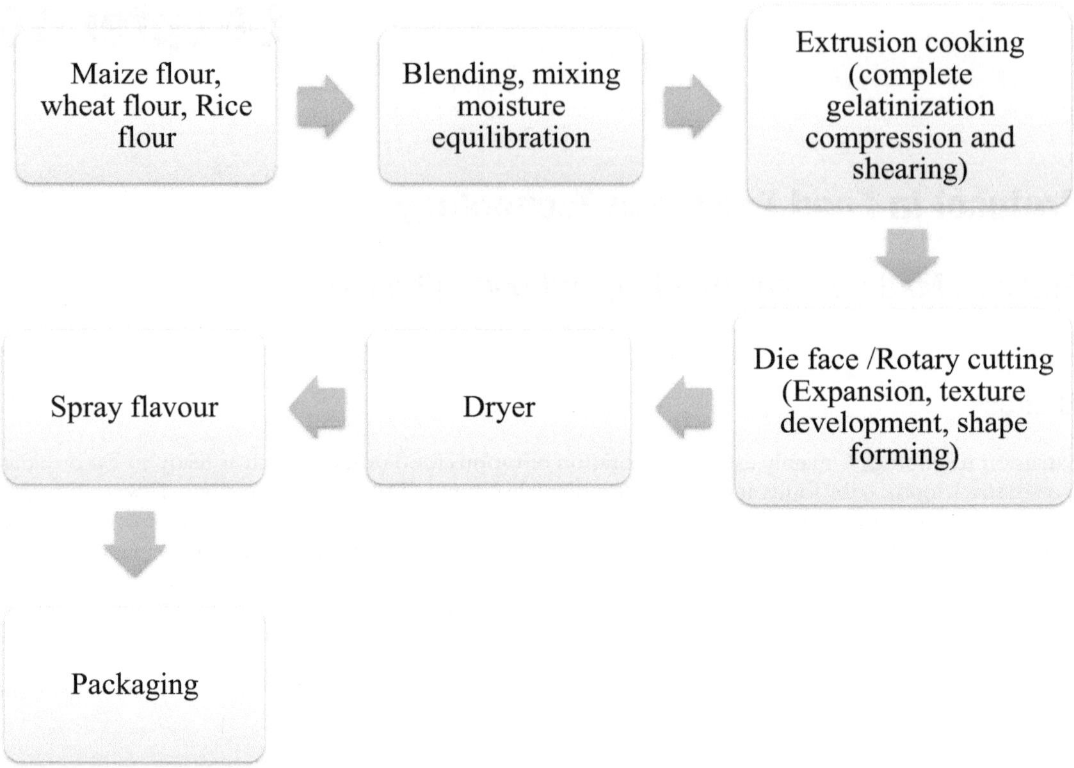

Fig. 1 Process flow chart of extrusion cooking

out. In the past, extrusion was used in metallurgy and processing of plastic polymers [8]. Extrusion cooking is a unique technology, which is used to process the raw materials and further used to develop different snack items using flour made of soy, corn, rice, wheat, potato, etc., into a desired shape and textures. The quality of the extruded product depends on the various parameters such as extruder type (i.e., screw, barrel temperature, screw speed, and feeding rate) and raw material (i.e., type of raw material, size, and moisture, etc.,) [9]. Additionally, food extrusion technology is used for pre-treatment in order to remove unwanted flavors, inactivate enzymes and antinutritional factors, modification of starch, and pre-gelatinization [7, 10].

Extruders are classified into various parts such as motor, screw, barrel, die, and knife/cutter which depends on mechanism [11]. Generally, food extruder is divided into single-screw and twin-screw extruders. In single-screw extruder, raw materials like maize grits, wholegrain, and rice cones are used which acts as good flow, that's why they are extensively used for extruding snacks and breakfast cereals [9]. The twin-screw extruder is widely categorized into two types, i.e., co-rotating and counter-rotating based on the relative motion of each screw to one another [12]. However, the

TSE has similar applications to the single-screw extruder, but TSEs can be used to carry out more challenging tasks, which the single-screw extruder may not be able to perform.

2 Materials

Extruded food products are prepared from various raw materials, which consist of different functional properties. During manufacturing and stabilization, it provides various color, nutritional components, and flavor. Extrusion cooking is a form of food processing, which involves high temperature, high pressure, and mechanical shearing, until it forms a final product [13]. Extrusion raw materials comprise flours made of a pea, rice, potato, barley, oats, cassava, tapioca, buckwheat, rye, and other grains. In preparation of TVP, raw materials used are proteinaceous materials such as soybean, wheat, sunflower seed, rape seed, fava beans, field beans, etc., and these food ingredients interact with each other and lead to transformation of foods [14]. Thereby, understanding the role of each ingredient is very important, as it may affect the final composition of extruded products.

3 Methods

3.1 Principle of Extrusion Technology

Extrusion involves series of steps, i.e., mixing, shearing, kneading, cooking, compressing under high pressure, and coming out through a die. Materials push out through high pressure and transform water into the steam. Thus, expansion/puffing of material takes place. Final product obtained is an extrudate, which usually depends on the shape of the die [15]. Apart from physiological change, denature of protein, fiber solubilization, and starch gelatinization take place during extrusion process. This shows a notable change in functional and chemical properties of an extruded material. Henceforth, extrusion is used extensively in production of snacks, pasta, frankfurters, and meat analogs [16].

Generally, extruder consists of a single or twin screw, which rotates in a cylindrical barrel, depending on the screw; these are classified into a single-screw extruder and a twin-screw extruder [15].

3.2 Single-Screw Extruder

A single-screw extruder consists of one rotating screw placed in a metal barrel, which can be seen in various patterns (i.e., screw configurations). The schematic diagram is shown in Fig. 2. The hollow cylindrical barrel is fitted tightly which heats food materials and gelatinizes during a continuous flow.

A single-screw cooking extruder consists of various parts such as:

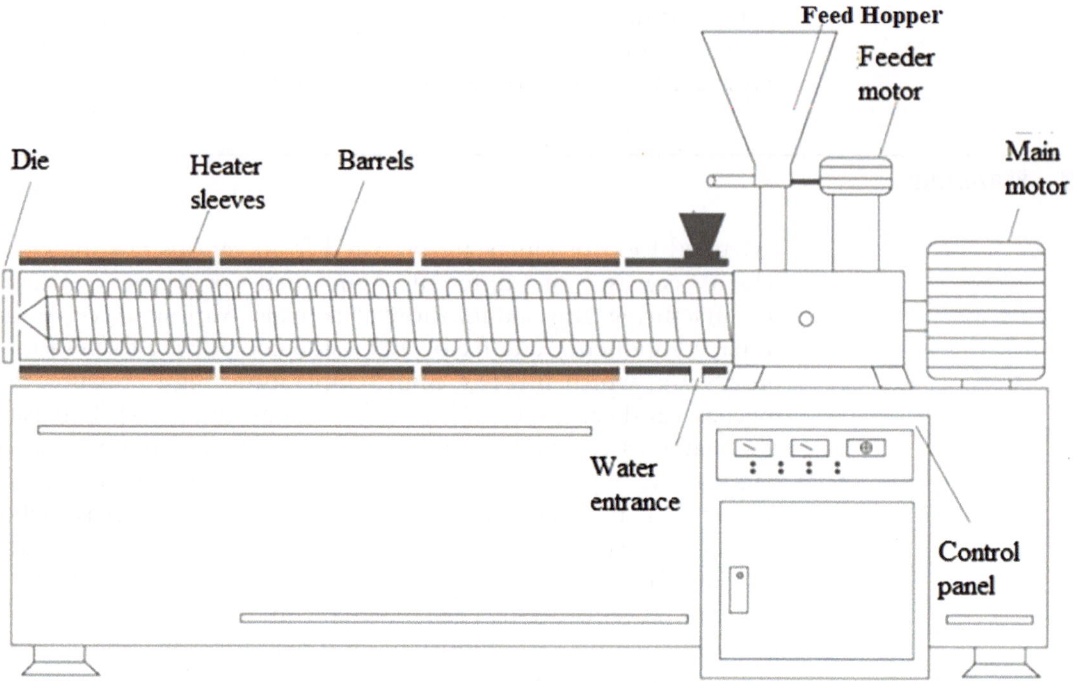

Fig. 2 Schematic diagram of a single screw extruder

1. Preconditioner: In this, most of the materials are mixed manually or by motor operated (*see* **Note 1**).
2. Feeder: The main purpose of this section is to deliver the raw materials to next part. Constant and uniform feeding of raw materials is necessary to undergo proper unit operations (*see* **Note 2**).
3. Barrel: It is main part of extruder system. Most of the work is done in this section. This part includes of screw, sleeves, barrel heads, and die. In this section, if it consists a single shaft, then it is a single-screw extruder, whereas if it consists two shafts, then it is a twin-screw extruder (*see* **Notes 3** and **4**).
4. Die: It is a part of extruder, which is placed at the end of the barrel (*see* **Note 5**).
5. Knife/cutter: Last part of the extrusion (*see* **Note 6**).

Extruders are generally classified by length-diameter (L/D) ratio and compression ratio. Generally, 3:1 is the most common compression ratio used during extrusion process.

3.3 Twin-Screw Extruder (TSE)

A TSE consists of two parallel screws, which is usually used for high moisture extrusion (more than 35%). The operation of a machine and its structure, behavior, and the outcome are entirely different from those of a single-screw extruder. The rotation of screw inside a TSE could generate the flow of material in eight-shaped form.

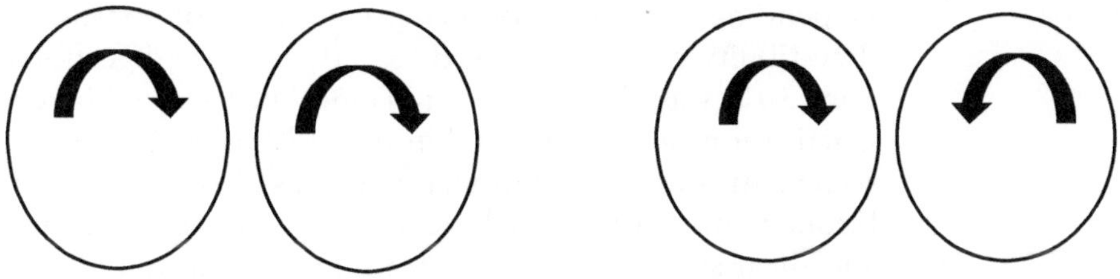

Fig. 3 Rotational direction of twin screw

Based on the screw position and its direction of rotation, it is classified into two types, i.e., co-rotating and counter-rotating (Fig. 3), in which co-rotating screw turns in the same direction and counter-rotating screw turns in the opposite direction.

A TSE consists of a preconditioning system, feeding section, screw, barrel, and die. Preconditioning is applied when raw materials consist of 20–30% moisture with longer residence time. It helps in reducing retention time and cost of energy. Feeding section allows uniform and constant feeding of raw materials for proper functioning of extruders. Screws are the main component of a TSE, which determines gelatinization, cooking temperature, protein denaturation, and starch dextrinization. Different combinations of screw elements could make the screws configuration for certain purpose. The barrel contains screws, and it allows movement of steam, water, or air which controls temperature of the TSE. The die gives different shapes to the end product. It can be in different design and shapes. A twin-screw extruder is complex and expensive compared to a single-screw extruder. It also poses several advantages over a single-screw extruder such as the TSE can handle oily, sticky, viscous materials which will not cause slippage. Broad range of particle size can be used in TSE [17].

4 Effect of Extrusion on Physicochemical Product Parameters

Color changes during extrusion are mainly because of non-enzymatic browning due to Maillard reaction among the proteins and sugars [18]. It occurs because of high barrel temperature with lower moisture which leads to the reduction of nutritional qualities [15]. In some cases, brittleness and crispness of a product increase due to increase in protein content which is negatively related with peak viscosity [19]. Extrusion leads to change in physicochemical components of food which is discussed below.

4.1 Protein

Since extrusion process involves higher temperature, denature of protein is common which impacts the structure of a protein.

Depending on the processing condition of extrusion, protein digestibility can be improved [5, 20]. Higher digestibility of protein (30%) was observed in canola meal because of higher degree of inactivation of protease inhibitors [21]. Also, extrusion inactivated some anti-nutritive components such as phytate and protease inhibitors. Similar study was done on pea protein in which temperature and shear stress showed significant impact on solubility [22].

4.2 Lipid

Lipids are usually present in lesser amount in extrudates, since it minimizes the friction needed to transfer heat energy. For instance, lipids act as plasticizer and also impart adhesive texture. Higher temperature would inactivate lipase and lipoxygenase activity and thus reduce the development and oxidation of fatty acids [23].

4.3 Dietary Fiber

Dietary fiber (total dietary fiber, TDF) is one of the major components in food product, since it is associated with reduction in heart-related diseases, cholesterol level, and colorectal cancer and controls gastrointestinal issues [24, 25]. Temperature, shear stress, and moisture content seriously affect TDF content [26]. Similarly, higher temperature as well as higher screw speed would enhance TDF in wheat bran extrudate products. This is mainly because of development of resistant starch at the time of storage [27]. Extrusion process improves soluble dietary fiber (SDF). A 10% SDF was found in extruded snack of soybean residue, and similar results were seen in lupin seed coat. Its SDF was increased from 3 to 9% after extrusion [28]. SDF content would be increased with increasing temperature and screw speed, decreasing moisture. Regardless, after certain time period, SDF would be stagnant or even tends to be decrease. This is because of depolymerization that occurs in polysaccharides [28]. These studies show that extrusion process could enhance or reduce TDF depending on the processing conditions.

4.4 Polyphenols and Antioxidants

These are naturally occurring components present in various fruits and vegetables, and various studies showed the change in phenolic content after extrusion. The combination of barley flour, tomato, and grape pomace resulted in decrease of phenolic content and antioxidant activity, which attribute to decomposition and evaporation at high temperature [29]. Similar results were found in corn flour in which phenolic content was reduced. While increase in phenolic acids such as ferulic and p-coumaric acid was also noted [30, 31].

Generally, antioxidant relies on both amount and composition of bioactive components. Accordingly, antioxidant activity is not affected by reduction in phenolic content. There was a significant enhancement in antioxidant activity and total phenolic content of germinated chickpea and tomato pomace-enriched snacks, i.e., 1.92–7.94% and 1.07–5.55%, respectively [32]. During extrusion cooking, antioxidant generally increases with temperature, whereas anthocyanin content gets reduced [33]. Furthermore, temperature and moisture tend to influence the compounds that are produced during Maillard reaction which directly affects antioxidant activity [34].

4.5 Antinutritional Factors

Antinutritional factors (ANFs) present in plants show negative effects on human health and also impact nutrient bioavailability. Common ANFs are present in food products such as saponin, oxalate, phytate, trypsin inhibitors, tannic acid, etc. Elimination of such ANFs during extrusion was studied [35]. A study found that preparing lentil and rice noodles at a higher temperature (180°C) with a higher moisture content (22%) reduced ANFs and enhanced nutritional bioavailability. Similar study reported that temperature above 140 °C and moisture at 20% showed ANFs can be removed successfully [36]. These studies showed screw speed, temperature, and moisture content are the main factors that influence the inactivation of anti-nutritive factors present in food during the extrusion process.

5 Application of Food Extrusion Technology

(a) Cold extrusion (*see* **Note** 7).

(b) Frozen extrusion (*see* **Note 8**).

(c) Co-extrusion (*see* **Note 9**).

(d) Reactive extrusion (*see* **Note 10**).

(e) Texturized vegetable protein (TVP), meat analog (*see* **Note 11**).

6 Mathematical Models Involved in Extrusion Processing

In extrusion cooking processes, various models have been proposed and mainly focused on the fundamental balances and prediction of steady and transient responses of various variables like change in power, feed rate, moisture, and screw speed. The ideal flow model for extruder and for residence time distribution (RTD) was studied [37]. Energy model, flow, and heurist models of extrusion were used during extrusion process [38].

7 Notes

The following notes (1–6) are related to the single-screw cooking extruder.

1. It is used to blend the raw materials into homogenous mixture with desirable hydration before being fed into the feeder, or it can be done manually.
2. It is continuously used to feed the mixture into the extruder constantly with proper consistency which is rotated by a motor. Generally, raw materials are feed gravitationally or volumetrically, and also, more than one feeder can be used in extrusion process if more than one ingredient got involved. Generally, the moisture content of feed is categorized into three levels, i.e., low moisture (up to 25%), intermediate moisture (between 25 and 50%), and high moisture feed (up to 50% and above) [39].
3. It consists of a hopper with a fitted screw which is rotated by a variable speed motor. Feeder supplies raw materials continuously into the extruder, and the barrel is heated externally using electricity. Raw materials are processed throughout the barrel and passed through the die into the desired shape. During the process, high temperature, shear force, moisture content, and screw speed modify raw materials, which are then reconstituted as a texturized product (i.e., texturized vegetable protein) protein. Also, moisture content and screw speed influence the final product. For example, in whole grain barley, 40% moisture with a screw speed at 50 rpm caused higher shear force, which resulted in a puffing texture, whereas at a lower screw speed with 30 rpm, it formed a texturized meat-like structure [40].
4. Barrels and heater sleeves: It is a part of the extruder, in which barrels are heated with the help of heater sleeves; this acts as heating jacket. Barrel temperature influences the conveying efficacy. This is due to change in friction factor among barrel liner and materials. Generally, the higher the surface temperature, the lesser the friction factor, which results in greater slippage. This is more common in single-screw than twin-screw extruders. As a result, higher barrel temperature influences higher torque and higher mechanical energy input [39, 41].
5. Die is an important part of the extruder, which acts as flow resistance to generate longer residence time of raw materials inside the barrel that helps to shape the final product [42].
6. The main function of a knife/cutter assembly is to cut the product into the desired shape and length, which is connected by a motor. Generally, the knife/cutter assembly includes different designs depending on the manufacturing company.
7. In cold extrusion, the optimum temperature is maintained for mixing and shaping foods such as pasta and meat products. Additional heat is not required, since it only shapes the food into different forms by applying pressure. It involves preparation of pasta, which is prepared from the wheat, semolina, and

other ingredients [43]. Ingredients are mixed and added into the extruder; it kneads the dough and pushes through the die to form a pasta. During cold extrusion, the processing conditions need to be maintained, i.e., lesser temperature makes higher dough viscosity, which makes dough difficult to pass through the die, whereas higher temperature along with increased screw speed improves the quality by decreasing the residence time of the final product [44]. Therefore, understanding the rheological behavior of dough is very important for pasta extrusion.

8. This process was used in the early 1990s for the preparation of ice cream. It is one of the recent techniques in food extrusion [45]. It is used in leading ice manufacturers which is said to be a low-temperature extrusion. The mixture is combined and used in conventional ice preparation. Controlling of shear and tear is maintained to form very small ice crystals, which results in smoother and creamier texture. Therefore, this process gives the best quality ice cream, and it forms good texture [46].
9. In this process, two or more materials are mixed simultaneously which have different texture, flavor, color, etc., to obtain a multiphase product. For example, a crispy cereal outer envelope can be co-extruded with a sweet or savory filling basis. Co-extrusion started in the 1990s in polymer industries, and later it was successfully tested in the food industry [47].
10. The reactive extrusion is a continuous mixing process at the molecular level, which is used for chemical and other biochemical reactions. This is a continuous extruder reactor which is suitable for the chemical industry for polymerization and its modification [48]. Furthermore, in the food industry, it is used in the modification and hydrolysis of starch. Reactive extrusion is highly suited to viscous media [49, 50].
11. TVP is a product made by a cooking extrusion process which uses vegetable proteins such as soy, legumes, and different kinds of nuts [51]. Furthermore, TVP is used in preparation of meat and poultry analogs which is an excellent substitute for meat. TVP can be classified into low-moisture meat substitute and high-moisture meat substitute. TVP possesses different characteristics such as light in weight, versatile texture, and high protein content with lesser percentage of fat and sodium [52]. TVP may vary based on the processing conditions such as the extruder, i.e., single or twin screw, diameter and length of screws, die, feed moisture content, feed rate, barrel temperature, and screw speed [53]. Moreover, a long cooling die is preferred to form a meat-like structure and avoid expansion. The characteristics of TVP are based on various specifications, i.e., protein content, oil content, particle size, sugar, emulsifying activity, and texturization agent.

8 Conclusion

Currently, extrusion technology has gained a lot of importance in the preparation of many food products including ready-to-eat snacks, meat analogs, baby foods, texturized vegetable protein and confectionary foods, etc. This chapter aims to understand the methods, principles, and protocols involved in extrusion processing technology. Extrusion processing has a great versatility in the development of highly nutritious food, with lesser cost which helps in replacing conventional methods of preparing snacks which have high processing cost and are less nutritious. Since extrusion process has been widely used these days, it has various beneficial effects such as increasing protein digestibility and total dietary fiber and reducing lipid oxidation, microbial contamination, and antinutritional factors which have a major role in the manufacture of a variety of extruded food products.

Nowadays, consumer demand for healthy foods such as protein-rich, high-fiber, gluten-free, and so on is growing, and many companies are investing in the development of novel food products, particularly plant-based proteins, which are closely linked to food extrusion technology. In the near future, updating the extruder system in order to make a production easier and faster is expected. In addition, we will see more applications of food extrusion technology not just in the underdeveloped countries but also in the developing ones.

References

1. Stojceska V, Ainsworth P, Plunkett A, İbanoğlu Ş (2009) The effect of extrusion cooking using different water feed rates on the quality of ready-to-eat snacks made from food by-products. Food Chem 114(1):226–232. https://doi.org/10.1016/j.foodchem.2008.09.043
2. de Cruz CR, Kamarudin MS, Saad CR, Ramezani-Fard E (2015) Effects of extruder die temperature on the physical properties of extruded fish pellets containing taro and broken rice starch. Anim Feed Sci Technol 199:137–145. https://doi.org/10.1016/j.anifeedsci.2014.11.010
3. Tiwari A, Jha S (2017) Extrusion cooking technology: principal mechanism and effect on direct expanded snacks–an overview. Int J Food Stud 6(1):113
4. Thymi S, Krokida MK, Pappa A, Maroulis ZB (2005) Structural properties of extruded corn starch. J Food Eng 68(4):519–526. https://doi.org/10.1016/j.jfoodeng.2004.07.002
5. Arêas J, Rocha-Olivieri C, Marques M (2016) Extrusion cooking: chemical and nutritional changes reference module in food science encyclopedia of food and health. Academic Press Elsevier, Amsterdam
6. Li M, Hasjim J, Xie F, Halley PJ, Gilbert RG (2014) Shear degradation of molecular, crystalline, and granular structures of starch during extrusion. Starch 66(7–8):595–605. https://doi.org/10.1002/star.201300201
7. Ding Q-B, Ainsworth P, Plunkett A, Tucker G, Marson H (2006) The effect of extrusion conditions on the functional and physical properties of wheat-based expanded snacks. J Food Eng 73(2):142–148
8. Berk Z (2017) Food extrusion. In: Engineering foods for bioactives stability and delivery. Springer, New York, pp 309–339
9. Frame ND, Harper JM (1994) The technology of extrusion cooking. Springer, New York
10. Lazou A, Krokida M (2010) Structural and textural characterization of corn–lentil extruded snacks. J Food Eng 100(3):392–408

11. Grandison AS, Brennan J (2012) Food processing handbook. Wiley-VCH, Chichester
12. Adekola K (1999) Studies into twin screw food extruder die. Unpublished PhD Thesis Changchun, China: Jilin University of Technology, Department of Agricultural Engineering, pp 108–134
13. Guy R, Horne A (1988) Extrusion cooking and co-extrusion. In: Food structure: its creation and evaluation. Butterworths, London, pp 331–349
14. Guy R (2001) 2 - Raw materials for extrusion cooking. In: Guy R (ed) Extrusion cooking. Woodhead Publishing, Sawston, pp 5–28. https://doi.org/10.1533/9781855736313.1.5
15. Fellows PJ (2009) Food processing technology: principles and practice. Elsevier, Amsterdam
16. Singh S, Gamlath S, Wakeling L (2007) Nutritional aspects of food extrusion: a review. Int J Food Sci Technol 42(8):916–929
17. Riaz MN (2019) Chap. 19 - Food extruders. In: Kutz M (ed) Handbook of farm, dairy and food machinery engineering, 3rd edn. Academic, Boston, pp 483–497. https://doi.org/10.1016/B978-0-12-814803-7.00019-1
18. Wang Y-Y, Ryu G-H (2013) Physicochemical and antioxidant properties of extruded corn grits with corn fiber by CO2 injection extrusion process. J Cereal Sci 58(1):110–116
19. Sharma C, Singh B, Hussain SZ, Sharma S (2017) Investigation of process and product parameters for physicochemical properties of rice and mung bean (Vigna radiata) flour based extruded snacks. J Food Sci Technol 54(6):1711–1720
20. Alonso R, Aguirre A, Marzo F (2000) Effects of extrusion and traditional processing methods on antinutrients and in vitro digestibility of protein and starch in faba and kidney beans. Food Chem 68(2):159–165
21. Zhang B, Liu G, Ying D, Sanguansri L, Augustin MA (2017) Effect of extrusion conditions on the physico-chemical properties and in vitro protein digestibility of canola meal. Food Res Int 100:658–664
22. Beck SM, Knoerzer K, Arcot J (2017) Effect of low moisture extrusion on a pea protein isolate's expansion, solubility, molecular weight distribution and secondary structure as determined by Fourier Transform Infrared Spectroscopy (FTIR). J Food Eng 214:166–174
23. Agarwal S, Chauhan ES (2019) Extrusion processing: the effect on nutrients and based products. Pharma Innov J 8(4):464–470
24. Aune D, Chan DS, Lau R, Vieira R, Greenwood DC, Kampman E, Norat T (2011) Dietary fibre, whole grains, and risk of colorectal cancer: systematic review and dose-response meta-analysis of prospective studies. BMJ 343:d6617
25. Threapleton DE, Greenwood DC, Evans CE, Cleghorn CL, Nykjaer C, Woodhead C, Cade JE, Gale CP, Burley VJ (2013) Dietary fibre intake and risk of cardiovascular disease: systematic review and meta-analysis. BMJ 347: f6879
26. Garcia-Amezquita LE, Tejada-Ortigoza V, Serna-Saldivar SO, Welti-Chanes J (2018) Dietary fiber concentrates from fruit and vegetable by-products: processing, modification, and application as functional ingredients. Food Bioproc Tech 11(8):1439–1463
27. Rashid S, Rakha A, Anjum FM, Ahmed W, Sohail M (2015) Effects of extrusion cooking on the dietary fibre content and Water Solubility Index of wheat bran extrudates. Int J Food Sci Technol 50(7):1533–1537
28. Zhong L, Fang Z, Wahlqvist ML, Hodgson JM, Johnson SK (2019) Extrusion cooking increases soluble dietary fibre of lupin seed coat. LWT 99:547–554
29. Altan A, McCarthy KL, Maskan M (2009) Effect of extrusion process on antioxidant activity, total phenolics and β-glucan content of extrudates developed from barley-fruit and vegetable by-products. Int J Food Sci Technol 44(6):1263–1271
30. Buitimea-Cantúa NE, Torres-Chávez PI, Ramírez-Wong B, Ledesma-Osuna AI, Gutiérrez-Uribe JA, Serna-Guerrero DA, Serna-Saldívar SO (2018) Ferulic, p-coumaric, diferulic and triferulic acids contents of corn tortillas prepared with extruded corn flour and enriched with sorghum (Sorghum bicolor (L.) Moench) bran. J Food Meas Charact 12(3): 1633–1640
31. Buitimea-Cantúa NE, Torres-Chávez PI, Ramírez-Wong B, Serna-Saldívar SO, Rouzaud-Sández O, Rosas-Burgos EC, Platt-Lucero LC, Salazar-García MG (2017) Phenolic compounds and antioxidant activity of extruded nixtamalized corn flour and tortillas enriched with sorghum bran. Cereal Chem 94(2):277–283
32. Hegazy HS, El-Bedawey A, Rahma E, Gaafar AM (2017) Effect of extrusion process on nutritional, functional properties and antioxidant activity of germinated chickpea incorporated corn extrudates. Am J Food Sci Nutr Res 4(1):59–66
33. White BL, Howard LR, Prior RL (2010) Polyphenolic composition and antioxidant capacity

of extruded cranberry pomace. J Agric Food Chem 58(7):4037–4042
34. Sharma P, Gujral HS, Singh B (2012) Antioxidant activity of barley as affected by extrusion cooking. Food Chem 131(4):1406–1413
35. Rathod RP, Annapure US (2017) Physicochemical properties, protein and starch digestibility of lentil based noodle prepared by using extrusion processing. LWT 80:121–130
36. Kaur S, Sharma S, Singh B, Dar B (2015) Effect of extrusion variables (temperature, moisture) on the antinutrient components of cereal brans. J Food Sci Technol 52(3): 1670–1676
37. Levine L (1982) Estimating output and power of food extruders 1. J Food Process Eng 6(1): 1–13
38. Cayot N, Bounie D, Baussart H (1991) Model identification in extrusion cooking. Food Control 2(3):140–145
39. Maskan M, Altan A (2011) Advances in food extrusion technology. CRC Press, Boca Raton
40. Chang C, Lin J (2017) Comparison between collet and cooking extrusions on physicochemical properties of whole grain barley. J Food Process Eng 40(3):e12480. https://doi.org/10.1111/jfpe.12480
41. Goff HD (2013) Chapter 9 - Dairy product processing equipment. In: Kutz M (ed) Handbook of farm, dairy and food machinery engineering, 2nd edn. Academic Press, San Diego, pp 199–221. https://doi.org/10.1016/B978-0-12-385881-8.00009-4
42. Ek P, Ganjyal GM (2020) Chapter 1 - Basics of extrusion processing. In: Ganjyal GM (ed) Extrusion cooking. Woodhead Publishing, Sawston, pp 1–28. https://doi.org/10.1016/B978-0-12-815360-4.00001-8
43. Kill RC, Turnbull K (2001) Pasta and semolina technology. Wiley Online Library
44. Abecassis J, Abbou R, Chaurand M, Morel M, Vernoux P (1994) Influence of Extrusion Conditions on Extrusion Speed, Temperature, and pressure in the extruder and on pasta quality. Cereal Chemistry 71(3):247–253
45. Eisner M (2006) Fat structure development in low temperature extruded ice cream, vol 26. ETH, Zurich
46. Roos YH, Livney YD (2017) Engineering foods for bioactives stability and delivery. Springer, New York
47. de Cindio B, Gabriele D, Pollini CM, Peressini D, Sensidoni A (2002) Filled snack production by coextrusion-cooking: 1. Rheological modelling of the process. J Food Eng 52(1):67–74
48. Mani R, Bhattacharya M, Tang J (1999) Functionalization of polyesters with maleic anhydride by reactive extrusion. J Polym Sci A Polym Chem 37(11):1693–1702
49. Linko P (1992) Twin-screw extrusion cooker as a bioreactor for starch processing. In: Food extrusion science and technology. Marcel Dekker, New York, pp 335–344
50. Xie F, Yu L, Liu H, Chen L (2006) Starch modification using reactive extrusion. Starch 58(3–4):131–139
51. Branch S, Maria S (2017) Evaluation of the functional properties of mung bean protein isolate for development of textured vegetable protein. Int Food Res J 24(4):1595–1605
52. Alamu EO, Busie M-D (2019) Effect of textured soy protein (TSP) inclusion on the sensory characteristics and acceptability of local dishes in Nigeria. Cogent Food Agric 5(1). https://doi.org/10.1080/23311932.2019.1671749
53. Zhang J, Ying D, Wei Y, Zhang B, Su X, Li S (2017) Thermal transition and decomposition properties of pH- and phosphate-induced defatted soybean meals. J Therm Anal Calorim 128(2):699–706. https://doi.org/10.1007/s10973-016-5991-8

Chapter 16

Instant Controlled Pressure Drop (DIC) as an Emerging Food Processing Technology

Joelle Nader, Tamara Allaf, and Karim Allaf

Abstract

Instant controlled pressure drop (DIC) technology is a thermomechanical process that enables several applications, such as microbiological decontamination, extraction by autovaporization of volatile active molecules, enhancement of drying (thus named swell-drying), and assisted solvent extraction of natural active molecules. In this chapter, we describe the technology, the type of equipment that is marketed, and the way it is being used and monitored. We also explain the applications that have been industrialized or developed at lab and semi-industrial scales.

Key words Instant controlled pressure drop DIC, Expansion, Texturing, Decontamination, Extraction

1 Introduction

Instant controlled pressure drop (DIC) is a thermomechanical process of great specificity and relevance in texturing, in microbiological decontamination, in extraction by autovaporization of volatile active molecules, and in its synergistic effect of combination with conventional processes, such as drying (thus named swell-drying), solvent extraction of natural active molecules (leading to the Tripolium operation), and oilseed pressing.

DIC was initially defined, designed, studied, and patented in its different application areas by Professor Allaf's team in 1988. The creation of ABCAR-DIC Process Company in France has allowed a leap forward in manufacturing and marketing the DIC machines. Thus, studies of scaling up in the concerned processes have led to the successful completion of their industrial development. The design of the machinery is being continuously improved in alignment with the emerging fourth industrial revolution. Thus, DIC is currently implemented in several industrial applications, mainly in

Mohsen Gavahian (ed.), *Emerging Food Processing Technologies*,
Methods and Protocols in Food Science, https://doi.org/10.1007/978-1-0716-2136-3_16,

the food, agroindustry, flavors, cosmetics, oils, and pharmaceutical sectors [1–5].

The general concept of DIC is based on subjecting the product to an abrupt pressure drop towards the vacuum (about 5 kPa) after a treatment at high temperature (up to 180 °C) during a short time period (5–60 s), depending on the desired outcome [6–9]. At first, the material is placed in a processing vessel (*see* **Note 1**) where a primary vacuum is established. This can ensure, in the next step, a more intimate contact between the saturated steam and the exchange surface [10]. Then, the product is exposed to a saturated steam pressure bringing up speedily, partially by convection and primarily by condensation, the product's exchange surface to the steam temperature; then, a conduction process will enable the heat transfer within the product. Subsequently, the abrupt pressure drop towards the vacuum induces an instantaneous autovaporization of water and other volatile molecules. This induces a similarly instantaneous cooling, which stops the thermal degradation [11–13]. The DIC treatment engenders bacterial decontamination of vegetative forms and spores, thus extending the shelf life of the final product.

Thermomechanical texturing has been studied on many plant-based materials. The processing temperature level and the initial moisture content of the product would define the viscoelastic and rheological behavior just before the relaxation. The instantaneous autovaporization generates an adequate amount of vapor acting as a source of mechanical expansion stress in the internal pores. As the pressure abruptly drops, the temperature level immediately drops to stabilize at the equilibrium level depending on the total pressure. The level of the vacuum pressure is defined according to the glass transition of the material. The speed of temperature drop implies maintaining the rheological viscoelastic behavior. Thus, the mechanical stresses produced by the generated vapor can trigger structural modification.

The amount of vapor instantly generated and released within the bacteria medium has a highly effective mechanical destruction effect especially in the case of spores. Decontamination can thus be achieved at relatively low temperature levels and short time.

The autovaporization of the volatile molecules can be carried out in successive cycles (multi-flash autovaporization). It leads to the definition of a highly efficient operation, especially used for the direct extraction of essential oils performed in few minutes (*see* **Note 2**). The short duration of the operation allows the preservation of the residual non-volatile compounds usually in an expanded matrix. The combination of successive DIC cycles led to the definition of a new solvent extraction operation called Tripolium. The subsequent extraction of these molecules can thus be intensified whatever solvent is used (water, ethanol, hexane, CO_2-SC, etc.).

The expansion of the initially compact structure of the plants also leads to an important intensification of the chemical reaction kinetics. Thus, processes carried out in an "in situ" mode can be defined to ensure the great improvement of biofuel production (biodiesel by transesterification, bio-jet fuel, etc.). For example, DIC texturing leads to an increase in oil extraction yield [14], while an increase in biodiesel production yield has been observed through the in situ transesterification process [15].

One of the first efforts of DIC applications was dedicated to the improvement of drying, especially of plant-based materials. Since the shrinkage and collapse phenomena significantly decrease the drying rate because of the large reduction in water diffusivity within the material, DIC texturing normally leads to a vigorous intensification of the final drying stage (*see* **Note 3**). Thus, especially because the DIC expansion process requires relatively low levels of moisture for many materials, DIC can be performed after an initial drying stage and be followed by a final conventional drying step (*see* **Note 4**). Thus, DIC texturing combined with conventional drying systems allows defining the innovative process called "swell-drying." It increases the drying kinetics, ensures a highly relevant microbiological decontamination, results in better preservation or even an improvement of the availability of active molecules, and usually reaches a texture appreciated by the consumer. Indeed, the final drying stage (*see* **Note 5**) from 20–30 to less than 5 g H_2O/g d.b. is achieved in a reduced time, often estimated as half the time required for a conventional drying technique. In addition, DIC has proven to be an effective technology that can significantly increase the expansion rate and improve the texture, color, sensory, and organoleptic properties (*see* **Note 6**). This was established for several products, including coffee beans, fruits, and vegetables such as potatoes, onions, carrots, etc. [16–18]. Besides the previously mentioned objectives, two of the most important application areas of the DIC process are pasteurization and sterilization. High-temperature/high steam pressure treatment for a short time followed by an instantaneous cooling following the pressure drop allows deactivation of enzymes. The thermal and mechanical impact leads to a remarkable 4-log reduction of spores and bacteria within seconds [8]. This makes the process much more efficient compared to a regular steam sterilization procedure [19, 20]. The swell-drying process reflects a green, energy-efficient technology that addresses the various issues of different conventional drying operations. Its application has been performed on many products [21–25] with numerous industrial realizations in the United States, Mexico, Canada, Spain, France, etc.

DIC autovaporization of volatile molecules has also been defined, studied, and optimized as a fast, efficient, and high-quality essential oil extraction technique [5, 19, 26–28]. The multi-flash autovaporization (MFA) involves successive DIC cycles intending

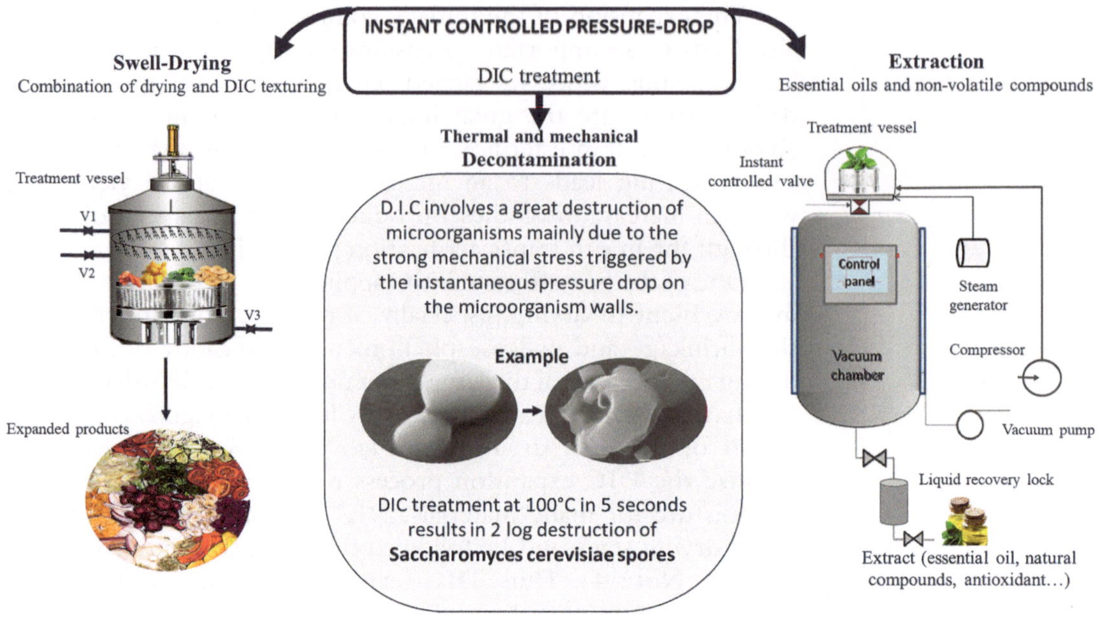

Fig. 1 DIC main areas of application

to evaporate the adequate molecules (*see* **Note 7**). Its efficiency is closely related to the molecule's volatility that is only a function of the exchange surface temperature. Since no possibility of intensification can be brought to the evaporation, MFA is distinguished by the absence of a paradoxical step, the presence of high kinetics, and a particularly reduced energy consumption (*see* **Note 8**). Thus, DIC-MFA leads to a significantly interesting extraction yield much higher than the steam distillation and hydrodistillation processes [29]. However, it is worth noting that the instantaneity of DIC autovaporization inevitably leads to a fog-like presence of the various essential oil (EO) compounds and water; thus, an EO water emulsion is obtained. Conventional evaporation or solvent separation step can be added as a way to achieve the separation between water and the EO molecules.

In brief, the main objectives and application areas of this innovative DIC technology are described in Fig. 1:

(a) Drying [11, 30, 31].

(b) Expansion [24, 32] and texturing for better water diffusion [33].

(c) Decontamination and destruction of microorganisms, spores, and fungi molds [34].

(d) Preservation and increase in the availability of the natural bioactive molecules [35].

(e) Elimination of allergic and antinutritional factors [36, 37].

(f) Improvement of digestibility.

(g) Extraction of natural compounds, such as antioxidants [6, 27, 35], polyphenols, and essential oils [38, 39].

(h) Processing of heat-sensitive products [35].

2 Materials

Instant controlled pressure drop (DIC) can therefore be adapted for multiple purposes (*see* **Note 9**), including texturing/expansion, drying, extraction, decontamination, direct extraction of volatiles, etc. Through the improvement of texturing and tortuosity, DIC increases the diffusivity of the solvents within the porous matrixes [40]. It also improves the accessibility and, thus, the extraction of non-volatile molecules. The DIC-expanded structure implies increasing the chemical and enzymatic reactions.

DIC is an ecological process that does not require the use of chemical agents. It is only based on thermomechanical mechanisms. It is worth mentioning that the same DIC equipment can be implemented for the different types and application areas of DIC. The only difference is normally related to the size and scale (laboratory and/or industrial scales) of equipment. The use of saturated steam at high pressure allows particularly fast kinetics of surface heating of the product. The time of thermal treatment by DIC is thus almost exclusively devoted to the homogenization of temperature and the moisture content within the material. The key point that distinguishes DIC from other conventional thermal processes is the abrupt pressure drop towards the vacuum that follows the short-term high-temperature phase (HTST). It results in instantaneous autovaporization/cooling. To ensure these controlled thermomechanical objectives, the following detailed equipment (Fig. 2) is required.

2.1 Processing Vessel

The processing vessel is the most important part of the DIC reactor where the material has to be processed. In a laboratory-scale device, it's a cylinder of about 18 cm diameter and 8 mm thick stainless steel screen basket. This acts as an autoclave where the HTST treatment takes place (up to 200 °C for a duration generally lower than 60 s) using high saturated steam pressure (up to 1000 kPa). The severity of the treatment conditions strongly depends on the process. The reactor, made of stainless steel, has evolved since the first machines in 1988 and up to now, to meet the specificities of each operation and product. It has also followed the ever-increasing mechanical and thermal constraints and the specific safety measures depending on the countries (USA, Japan, China, etc.). The automation and continuous monitoring of the operating parameters of temperature, pressure, and duration of the operation have greatly

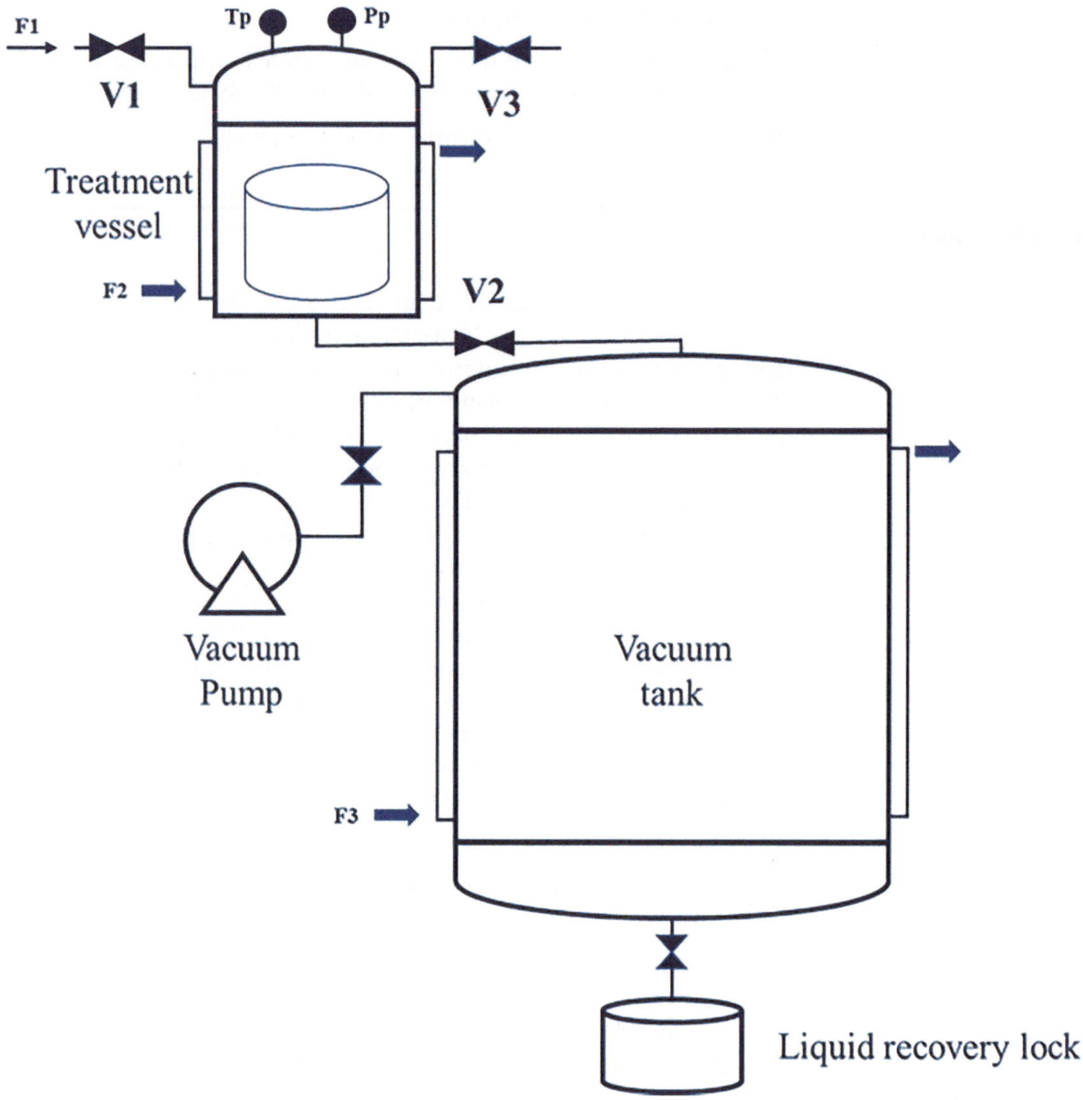

Fig. 2 Schematic diagram of DIC. *V1, V2, V3* valves, *F1-F2* saturated steam injection, a double jacketed vacuum tank with cooling water (F3), *Tp* Thermocouples, *Pp* Pressure gauge

modified the design (*see* **Notes 10** and **11**). Various shapes have also been adopted (cylindrical, elliptical), in vertical, horizontal, or inclined positions, with an average internal volume ranging from 3 to 12 liters (laboratory equipment) or from 1.5 to 2 m^3 (industrial equipment). The processing vessel is systematically equipped with a double heating jacket that partially covers the vessel surface to better maintain thermal uniformity in volume and time.

- Material: 304 stainless steel.
- Treatment conditions: different saturated steam pressure conditions can be provided depending on the nature of the product to be treated and the purpose of the treatment.

2.2 Connection System

The abrupt pressure drop towards the vacuum is provided by a large diameter valve connecting the processing vessel to the vacuum tank. This valve must be opened in a very short time (in less than 0.05 s) to ensure a controlled and rapid release of the steam pressure, initially injected into the processing vessel, to the vacuum system. The large-volume vacuum tank is connected to an adequate vacuum pump. An airlock system provides discharge while maintaining the vacuum within the tank. This is an electro-pneumatic ball valve with a diameter of 20 cm that allows performing the pressure drop towards a vacuum in a time of about 100 ms (*see* **Note 12**).

Another valve connects the steam generator or steam buffer tank to the processing vessel and controls the pressure of the saturated steam in the DIC reactor.

2.2.1 Connection Types

Connection site	Connection type
Steam generator and processing vessel	304 stainless steel piping
Compressor and DIC vessel	Adequate piping
DIC and vacuum tank	Stainless steel piping
Vacuum tank and vacuum pump	304 stainless steel piping
DIC and vacuum pump	Power supply
Airlock system and DIC	Electrical and pneumatic supply
Internet connection	Wireless
Electrical power supply	400 V/3 phases and ground, depending on the installation place

2.3 Vacuum System

The vacuum system consists of a large vacuum tank, whose volume (1–20 m^3) is 75–150 times greater than the volume of the treatment container, connected to an adequate vacuum pump (2.5–18 kW liquid ring vacuum pump). Usually, the liquid used is water whose temperature is kept below 18–20 °C to maintain a constant vacuum level of 3.5–5 kPa during all experiments. The vacuum tank is equipped with a cooling jacket that helps to establish the vacuum in the tank by continuously circulating a cooling fluid (15–18 °C); the energy losses for maintaining the vacuum level by condensation of a part of vapor are then drastically reduced, and the minimum vacuum level can reach 2.5 kPa. The vacuum pump is mounted on a skid with forklift pockets and lifting wedges for advanced mobility on-site.

- Material: 304 stainless steel.
- Vacuum tank volume: 1–2 m^3 for laboratory scale and 3–20 m^3 for industrial scale.

2.4 Airlock System

An automatic airlock system ensures an automatic draining of the vacuum tank manufactured with stainless steel piping. Thus, the recuperated solution of water and volatile compounds obtained by autovaporization or non-volatile mixtures obtained by expulsion can then be adequately valorized.

- Material: stainless steel 304.

2.5 Steam Generator

A food-grade steam generator is used. It is connected to the processing vessel through a butterfly valve to inject saturated steam up to 10 bar.

- Steam flow rate of 30 kg/h at laboratory scale and 100 kg/h at industrial scale.
- Steam drying 99%.
- Always use the steam generator with a steam reserve and a water droplet trap with care to not directly inject the steam coming from the steam generator to the processing vessel.

2.6 Air Compressor and Airflow Generator

- Pressure: 10–15 bars.
- Air compressor capacity: 250–750 L/min.

2.7 Cooling System

The cooling system is defined for the double jacket of the vacuum tank and possibly for getting a closed water circuit for the vacuum pump.

- Capacity: 2.5–18 kW at 15 °C.

2.8 Extract Container

Extract and condensate liquids can be collected from the vacuum vessel through the airlock chamber.

- Material: stainless steel 304.

2.9 Footprint

2.9.1 DIC Device

The footprint is about (length × width × height) 812 × 1285 × 2050 mm for laboratory scale and 4000 × 2600 × 2200 mm for industrial scale.

2.9.2 Vacuum Tank

- Volume: 1–3 m^3 at laboratory scale and 3–20 m^3 at industrial scale.
- Diameter: 1250 mm at laboratory scale and 1500 mm at industrial scale.

– Height: 2050 mm at laboratory scale and 2700 mm at industrial scale.

2.9.3 Pump

– Length × width × height: 650 × 675 × 495 mm at laboratory scale and 900 × 900 × 800 mm at industrial scale.

These dimensions are averages and may vary slightly. Depending on the size, options, and facilities, the layout can be adapted to find the best compromise between functionality, ease of use, and facility suitability.

2.10 Data Logging Software

Data logging software is installed on a computer with a touch screen that is connected to the DIC unit. The following objectives are then achieved:

– Display of all parameters.
– Selection and recording of process parameters.
– The DIC as a critical control point: Triggering of alarms in case of defective components. The cycles are analyzed in real time, and any processing error gives rise to an alert so that operators can discard the product that has not been processed correctly.
– The data is securely stored in the ABCAR Cloud for 3 years, and reprocessed, comparing the results with other cycles performed over the years.
– Results are accessible via a dedicated website and can be exported.
– The data analysis is also used for preventive maintenance.

3 Methods

Figure 3 illustrates the successive steps of (A) a single-cycle DIC process and (B) a multi-cycle process. A comprehensive overview of each step with its objectives and underlying theories is detailed below.

3.1 Extraction of Volatile and Non-volatile Compounds

3.1.1 Define the Operating Parameters

The main processing parameters are the saturated steam pressure and the processing time. These parameters may change depending on the nature of the product treated (*see* **Notes 13** and **14**) and the main objective of the DIC treatment adopted. At both laboratory and industrial scales, we can adopt the same optimized values of the processing parameters (*see* **Note 15**).

3.1.2 Place the Product in a Processing Vessel

We start by placing and loading the product in the suited baskets. Inside the treatment vessel, there is atmospheric pressure, and the reactor doors are open. On the industrial scale, the baskets are

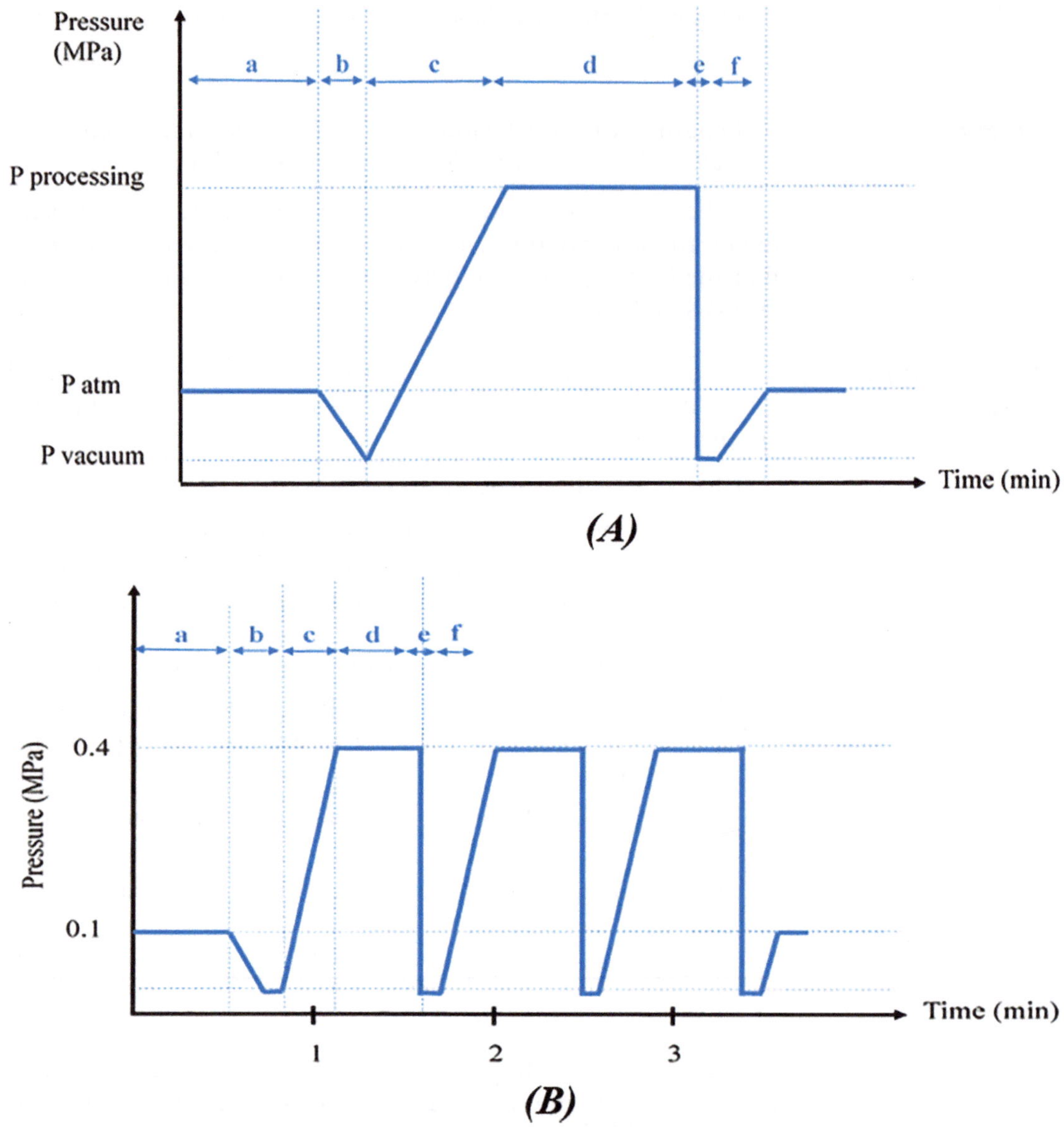

Fig. 3 (**A**) DIC treatment [23]; (**B**) Multi-cycle DIC [19]. (*a*) Treatment vessel at atmospheric pressure; (*b*) initial vacuum of 5 kPa; (*c*) injection of saturated steam; (*d*) maintaining the pressure for a specific time; (*e*) abrupt pressure drop; (*f*) vacuum period followed by restoration of atmospheric pressure

placed on the conveyor that will automatically transport them to the processing chamber and close the processing vessel doors.

3.1.3 Apply a First Vacuum Step (5 kPa) Just before Injecting the Steam

This first step allows closer thermal contact between the steam (i.e., the heating fluid) and the product exchange surface. Definitely, the open pores of the matrix are usually filled with air at atmospheric pressure. The latter is comparable to an insulating medium, thereby decreasing the thermal conductivity and bypassing the direct contact with the condensed water. Thus, the initial vacuum is a crucial step that facilitates the passage of the steam deeper within the product thickness and triggers a rapid increase in heat transfer.

3.1.4 Connect the Steam Generator to the Treatment Vessel

Open the ball valve (V1) connecting the steam generator to the treatment vessel to inject saturated steam with a pressure (P) varying between 5 and 1000 kPa and a temperature ranging from ambient temperature to 180 °C (according to preliminary studies). When the desired pressure P is reached, close the valve (V1).

3.1.5 Maintain the Product at High Saturated Vapor Pressure for a Short, Predetermined Period for Thermal and Moisture Homogenization

The pressure inside the treatment vessel is maintained constant for a short time to homogenize the temperature and water content inside the product. The saturated vapor pressure is usually maintained constant at a value in the range between 50 kPa and usually 700 kPa, for a short, predetermined period (5–60 s). The temperature varies depending on the pressure range selected but is typically between 80 and 160 °C. This "processing time" is necessary for the product to reach thermal and moisture equilibrium. During this time, the product acquires the desired rheological viscoelastic properties.

3.1.6 Instantaneously Open the Main Vacuum Valve

When the HTST treatment is finished, turn off the steam and abruptly open the main vacuum valve (V2) to instantaneously (in less than 50 or 100 ms) connect the treatment vessel to the vacuum tank. An instant pressure drop towards the vacuum, at a rate ($\Delta P/\Delta t$) greater than 0.5 MPa/s, is performed in the processing vessel. This event induces instantaneous autovaporization of water and volatile compounds present in the solid matrix and a rapid cooling of the product. This blocks further thermal degradation and ensures better extract quality [16]. Depending on the treatment condition, expansion, swelling, and/or texturing of the porous material normally occur. For higher severity of the treatment, the cell walls can break down, resulting in increasing the availability of metabolites. Due to the instantaneity of both decompression and temperature drops, the amount of vapor produced would trigger mechanical stress without using extremely high processing temperatures [18, 41] with a higher probability of reaching the glass transition. Consequently, the porous structure and honeycomb texture generated by autovaporization coupled with a high total pressure gradient between the interior (high central vapor pressure) and the exterior (vacuum) surface of the treated product (Darcy-type law) enhance the mass transfer of the targeted compounds (essential oils and others), allowing for a higher effective diffusivity and intensifying the overall kinetics.

3.1.7 Injection of an Airflow to Reach the Atmospheric Level

Through the opening of the valve V3, airflow can be spread and well distributed on the whole product. This relaxation from the atmospheric pressure towards the vacuum induces a further cooling of the product to better approach the glass transition.

3.1.8 Recovering of the Mixture from the Vacuum Tank

After collecting the products from the processing vessel, the extracts and liquids can be recovered from the vacuum tank. At the end of each DIC compression/decompression cycle, the volatile compounds pass through the valve into the vacuum tank where condensation of the components occurs through the cooling water jacket. The mixture of water, condensed volatile compounds, and expelled non-volatile molecules is collected from the liquid recovery lock, which is connected to the vacuum tank.

3.1.9 Multi-DIC Cycles

In some cases, multiple DIC cycles may be performed (Fig. 3B). In this case, the final step of atmospheric pressure recovery (g) is replaced by step (c) of saturated steam injection to start a new DIC cycle. The multi-cycle DIC consists of n repetitions of steps (c–f), where the total heating time is the sum of several heating times of the subsequent DIC cycles. The number of repetitions should be specified based on preliminary studies. A multi-cycle pressure drop is sometimes adopted to amplify the effect of the process, more specifically for thermolabile products, or to achieve more effective decontamination.

3.1.10 Complementary Operations

A final liquid-liquid extraction step is often necessary to separate volatile compounds or essential oils from the collected oil-in-water emulsion. Otherwise, the appropriate DIC extraction process does not require the use of organic solvent.

3.1.11 Energy Consumption

In general, the energy consumption per cycle has been estimated to be between 0.11 and 0.5 kWh per kg of raw material [11].

3.2 Drying, Texturing, Expanding, and/or Decontamination

3.2.1 Preparation Stage of the Product

Initial dehydration or partial drying of the product is necessary as a pretreatment to ensure a lower and more homogeneous water content. For other products, rehydration is needed to reach a viscoelastic behavior of the product at high temperature.

3.2.2 Introduce the Product into the Processing Vessel

The product is placed in the processing vessel where an initial vacuum (about 30 mbar) is applied (*see* **Note 16**). The purpose of this step is to facilitate and promote close heat (by condensation) exchange between the injected saturated steam and the product surface.

3.2.3 Thermal Treatment

The product is then subjected to a high saturated steam pressure (between 0.1 and 0.6 MPa) maintained for a short treatment period (less than 50 s). Initially, the condensation of the steam provides a high heat transfer coefficient that allows the surface of the product to be heated. At this stage, the HTST treatment is essential to allow the diffusion of condensed water from the surface to the core of the matrix and thus homogenize the temperature and water content of the product before the pressure drops.

3.2.4 Instant Pressure Drop

Once the water content and temperature levels are nearly homogenized in the material, the sample is subjected to an instant controlled pressure drop DIC (with a decompression rate $\Delta P/\Delta t$ greater than 0.5 MPa/s) towards a vacuum (4–5 kPa) which can cause a change in the product structure (such as expansion, increasing of tortuosity, etc.). The vacuum is usually maintained for 5–20 s.

This thermo-hydro-mechanical process causes a rapid autovaporization of part of the water contained in the product, which induces a controlled texturing of the treated sample, coupled to an instant cooling that stops any thermal degradation. Indeed, the difference between the internal pressure caused by the autovaporization and the external (vacuum) pressure results in mechanical stress within the matrix pores. This texturing process is therefore generally achieved when the water content almost reaches the glass transition phase of the material.

3.2.5 Establish the Atmospheric Pressure

The release of pressure in the processing vessel from the vacuum step to the atmospheric pressure may be carried out by spreading an airflow that can be well distributed on the whole product. This allows a better cooling of the product.

3.2.6 Collect the Treated Product from the Treatment Vessel

After completing the DIC treatment, place the treated products in a dryer to reach the desired moisture content (usually close to 5 g H_2O/g d.b. depending on the product composition) to obtain a dried product. A static bed dryer can be used with a 2 cm thin layer of DIC-treated sample that is subjected to an airflow at a velocity greater than the critical airflow velocity (CAV) and a constant inlet temperature of about 50 °C. The drying behavior of the DIC expanded, textured, and decontaminated product strongly depends on the hydro-thermo-rheological attributes of the treated product.

DIC treatment of partially dried material should be followed by a freezing stage (dehydro freezing process).

4 Notes

1. Disperse the product on the surface of the treatment vessel allowing better heat transfer to the core of the material matrix.
2. In the case of essential oil extraction, be sure to extend the duration of each cycle for a period longer than the critical heating time needed to ensure uniform distribution of temperature and moisture content throughout the material, which is 5–15 s. Then, the vacuum time must be sufficient to allow the removal of essential oil vapor according to Darcy's law, which is 3–10 s.
3. In the case of food drying, it is strongly recommended to control the DIC processing parameters to avoid significant losses of flavors and other volatile compounds, changes in texture and color, and a decrease in dietary value.
4. The product should be dried before and/or after the DIC treatment to sufficiently reduce the moisture content.
5. Drying methods that fundamentally overcome the inconsistency associated with conventional airflow drying are generally recommended after DIC processing. These include techniques that allow for a uniform heating process in depth, such as microwave treatment, or methods based on Darcy's vapor transfer law. In these cases, free or bound residual water is easily removed from the porous matrix due to the pressure gradient remaining the driving force.
6. DIC dried products can be easily ground, and the resulting expanded powders are characterized by uniform particle size, high specific surface area, improved organoleptic quality, and prodigious functional properties.
7. The industrial DIC unit is designed to process up to 60 cycles per hour (based on a 15 s DIC steam cycle). For fruit and vegetable densities, the unit will process between 75 and 150 kg/h. For seeds, legumes, or beans, the unit will process approximately 400 kg/h.
8. DIC treatment has proven to be an energy-efficient, cost-effective, and environmentally friendly process.
9. Depending on your application and the use of the equipment, the level of steam pressure, compressed air pressure, and cooling power may be different. The amount of steam depends on the product being processed. These aspects will be defined to suit your application.
10. Be sure to comply with international safety standards, especially for pressure vessels and electrical devices. The equipment can be designed and built, for example, in accordance with the

European Union safety standard for machine construction (CE conformity).

11. The equipment must have several independent and redundant safety levels on all critical parameters. The control system must always put the devices in a passive state in case of alarm or fault detection.
12. Ensure that all components of the experimental setup are well and tightly connected and that no leakage would occur during the process.
13. Careful selection of optimal operating parameters is of great importance, especially when processing heat-sensitive products such as bioactive phytochemicals and thus avoiding their denaturation during drying.
14. The severity of DIC treatment must be controlled as it can preserve, disrupt, shrink, or even break down the cell walls of the treated product.
15. To achieve the highest possible efficiency of a DIC process, various controlled operating parameters must be optimized, including the initial temperature and pressure, input steam pressure, initial vacuum pressure, saturated steam pressure, treatment time, pressure/temperature drop rate, number of DIC cycles, a ratio of vacuum tank volume to treatment vessel volume, apparent volume of the treatment vessel, amount of product to be treated, etc. Besides, certain intrinsic characteristics must be controlled and optimized: product water content, raw material size and shape, effective thermal and mass diffusivity, specific heat capacity, thermal conductivity, and rheological properties such as glass transition, viscosity, elasticity, etc.
16. The initial vacuum step can be achieved by direct pumping (new connection with a vacuum pump) rather than by opening the valve V2 connecting the processing vessel to the vacuum tank. The use of mechanical energy is then less expensive and more cost-effective.

Acknowledgments

The authors thank ABCAR-DIC PROCESS SAS (Compiègne, France) for providing information on laboratory, pilot, and industrial scale DIC units.

References

1. Rakotozafy (2001) Application du nouveau procédé de déshydratation par détentes successives (DDS), dans le séchage de produits biologiques à haute valeur ajoutée. Dissertation, University of La Rochelle
2. Allaf K, Louka N, Bouvier JM, Parent F, Forget M (1993) Procédé de traitement de produits biologiques en vue de la modification de leur texture, installations pour la mise en œuvre d'un tel procédé et produits ainsi réalisés. French patent FR 93/09720, Extension Internationale No. PCT/FR94/00975
3. Allaf K, Rezzoug SA, Cioffi F, Contento MP (1998) Processus de Traitement Thermo-Mécanique par Détente Instantanée Contrôlée des Fruits, Jus et Peaux d'Agrumes. French patent no. 98/11105
4. Allaf T, Tomao V, Besombes C, Chemat F (2013) Thermal and mechanical intensification of essential oil extraction from orange peel via instant autovaporization. Chem Eng Process Process Intens 72:24–30. https://doi.org/10.1016/j.cep.2013.06.005
5. Chemat F, Rombaut N, Meullemiestre A, Turk M, Perino S, Fabiano-Tixier A-S, Abert-Vian M (2017) Review of green food processing techniques. Preservation, transformation, and extraction. Innov Food Sci Emerg Technol 41:357–377. https://doi.org/10.1016/j.ifset.2017.04.016
6. Chemat F, Abert Vian M, Fabiano-Tixier A-S, Nutrizio M, Režek Jambrak A, Munekata PES, Lorenzo JM, Barba FJ, Binello A, Cravotto G (2020) A review of sustainable and intensified techniques for extraction of food and natural products. Green Chem 22(8):2325–2353. https://doi.org/10.1039/C9GC03878G
7. Messaoudi Y, Smichi N, Allaf T, Allaf K, Gargouri M (2015) Effect of instant controlled pressure drop pretreatment of lignocellulosic wastes on enzymatic saccharification and ethanol production. Ind Crop Prod 77:910–919. https://doi.org/10.1016/j.indcrop.2015.09.074
8. Mounir S, Allaf T, Mujumdar AS, Allaf K (2012) Swell drying: coupling instant controlled pressure drop DIC to standard convection drying processes to intensify transfer phenomena and improve quality—an overview. Drying Technol 30(14):1508–1531. https://doi.org/10.1080/07373937.2012.693145
9. Téllez-Pérez C, Cardador-Martínez A, Mounir S, Montejano-Gaitán JG, Sobolik V, Allaf K (2013) Effect of instant controlled pressure drop process coupled to drying and freezing on antioxidant activity of green "Poblano" Pepper (Capsicum annuum L.). Food Nutr Sci 04(03):321–334. https://doi.org/10.4236/fns.2013.43043
10. Bouallegue K, Allaf T, Besombes C, Younes RB, Allaf K (2019) Phenomenological modeling and intensification of texturing/grinding-assisted solvent oil extraction: case of date seeds (Phoenix dactylifera L.). Arab J Chem 12(8): 2398–2410. https://doi.org/10.1016/j.arabjc.2015.03.014
11. Allaf T, Allaf K (eds) (2014) Instant Controlled Pressure Drop (D.I.C.) in food processing: from fundamental to industrial applications. Food engineering series. Springer, New York
12. Bouallegue K, Allaf T, Ben Younes R, Téllez-Pérez C, Besombes C, Allaf K (2020) Pressure, temperature and processing time in enhancing Camelina sativa oil extraction by Instant Controlled Pressure-Drop (DIC) texturing pre-treatment. Grasas Aceites 71(3):365. https://doi.org/10.3989/gya.0337191
13. Hernandez-Aguirre AI, Téllez-Pérez C, San Martín-Azócar A, Cardador-Martínez A (2019) Effect of instant controlled pressure-drop (DIC), cooking and germination on non-nutritional factors of common Vetch (Vicia sativa spp.). Molecules 25(1):151. https://doi.org/10.3390/molecules25010151
14. Jablaoui C, Besombes C, Jamoussi B, Rhazi L, Allaf K (2020) Comparison of expander and Instant Controlled Pressure-Drop DIC technologies as thermomechanical pretreatments in enhancing solvent extraction of vegetal soybean oil. Arab J Chem 13(10):7235–7246. https://doi.org/10.1016/j.arabjc.2020.08.005
15. Bamerni F, Kamal I, Allaf K (2017) Swell-texturing assisted in-situ transesterification of Camelina seeds biodiesel. Int J Eng Res Dev 13(9):31–44
16. Haddad J (2002) Impacts des procédés hydro-thermo-mécaniques dans la valorisation des graines oléoprotéagineuses. Dissertation, University of La Rochelle
17. Kamal IM, Sobolik V, Kristiawan M, Mounir SM, Allaf K (2008) Structure expansion of green coffee beans using instantaneous controlled pressure drop process. Innov Food Sci Emerg Technol 9(4):534–541. https://doi.org/10.1016/j.ifset.2008.01.004
18. Louka N, Allaf K (2004) Expansion ratio and color improvement of dried vegetables texturized by a new process "Controlled Sudden Decompression to the vacuum". J Food Eng

65(2):233–243. https://doi.org/10.1016/j.jfoodeng.2004.01.020

19. Kristiawan M, Sobolik V, Allaf K (2008) Isolation of Indonesian cananga oil using multicycle pressure drop process. J Chromatogr A 1192(2):306–318. https://doi.org/10.1016/j.chroma.2008.03.068
20. Setyopratomo P, Fatmawati A, Savitri E, Sutrisna PD, Allaf K (2019) Impact of instant-controlled pressure drop treatment on thermal properties and microbial decontamination of banana flour. AIP Conf Proc 2114(1): 050014. https://doi.org/10.1063/1.5112458
21. Mounir S, Ghandour A, Téllez-Pérez C, Aly AA, Mujumdar AS, Allaf K (2020) Phytochemicals, chlorophyll pigments, antioxidant activity, relative expansion ratio, and microstructure of dried okra pods: Swell-drying by instant controlled pressure drop versus conventional shade drying. Drying Technol 39:2145–2159. https://doi.org/10.1080/07373937.2020.1756843
22. Louka N, Juhel F, Allaf K (2004) Quality studies on various types of partially dried vegetables texturized by Controlled Sudden Decompression. J Food Eng 65(2):245–253. https://doi.org/10.1016/j.jfoodeng.2004.01.021
23. Mazen Hamoud-Agha M, Allaf K (2020) Instant controlled pressure drop (DIC) technology in food preservation: fundamental and industrial applications. In: Socaci SA, Farcas AC, Aussenac T, Laguerre J-C (eds) Food preservation and waste exploitation. IntechOpen, London. https://doi.org/10.5772/intechopen.83439
24. Mounir S (2007) Studies of new manufacturing process of powders by inserting the instant controlled pressure drop DIC within spray-drying, swell-drying and controlled vacuum atomization of dairy products. Dissertation, University of La Rochelle
25. Albitar N, Mounir S, Besombes C, Allaf K (2011) Improving the drying of onion using the instant controlled pressure drop technology. Drying Technol 29(9):993–1001. https://doi.org/10.1080/07373937.2010.507912
26. Stratakos AC, Koidis A (2016) Methods for extracting essential oils. In: Essential oils in food preservation, flavor and safety. Elsevier, Amsterdam, pp 31–38
27. Berka-Zougali B, Hassani A, Besombes C, Allaf K (2010) Extraction of essential oils from Algerian myrtle leaves by instant controlled pressure-drop D.I.C. technology. J Chromatogr A 1217(40):6134–6142. https://doi.org/10.1016/j.chroma.2010.07.080
28. Meullemiestre A, Breil C, Abert-Vian M, Chemat F (2015) Innovative techniques and alternative solvents for extraction of microbial oils. In: Meullemiestre A, Breil C, Abert-Vian M, Chemat F (eds) Modern techniques and solvents for the extraction of microbial oils. Springer, Berlin, pp 19–42
29. Chemat F, Vorobiev E (2020) Green food processing techniques: preservation, transformation and extraction. Academic, Boston. https://doi.org/10.1016/C2017-0-03313-X
30. Sahyoun W (1996) Maîtrise de l'aptitude de matériaux agro-alimentaires aux procédés de séchage. Etude de l'adéquation entre les états structuraux, biochimiques, physiques et comportementaux sur les processus de déshydratation = Mastering the aptitude of food products to drying processes. Study of the adequacy between structural, biochemical, physical, and behavioural states on the dehydratation process. Dissertation, Université de Technologie de Compiègne
31. Juhel F (2000) Etude de l'application de deux nouvelles opérations de séchage/texturation du poisson; analyse comparative du procédé et du produit. Dissertation, University of La Rochelle
32. Allaf K (2002) Analysis of instantaneity in thermodynamic processes; fundamental laws. Paper presented at the first Franco-Lebanese symposium on technologies and studies on process engineering and biochemistry, Beyrouth, 22–25 July 2002
33. Ben Amor B (2008) Maîtrise de l'aptiture technologique de la matière végétale dans les opérations d'extraction de principes actifs; texturation par Détente Instantanée Contrôlée DIC. Dissertation, University of La Rochelle.
34. Albitar N, Mounir S, Besombes C, Allaf K (2011) Improving the drying of onion using the instant controlled pressure drop technology. Drying Technol 29(9):993–1001. https://doi.org/10.1080/07373937.2010.507912
35. Mounir S, Besombes C, Al-Bitar N, Allaf K (2011) Study of instant controlled pressure drop DIC—treatment in manufacturing snack and expanded granule powder of apple and onion. Drying Technol 29(3):331
36. Haddad J, Allaf K (2007) A study of the impact of instantaneous controlled pressure drop on the trypsin inhibitors of soybean. J Food Eng 79(1):353–357. https://doi.org/10.1016/j.jfoodeng.2006.01.066
37. Haddad J, Greiner R, Allaf K (2007) Effect of instantaneous controlled pressure drop on the phytate content of lupin. LWT Food Sci Technol 40(3):448–453

38. Allaf K (2009) The new instant controlled pressure—drop DIC technology. In: Chemat F (ed) Essential oils and aromas: green extraction and application. Har Krishan Bhalla and Sons, New Delhi, pp 85–121

39. Besombes C, Berka-Zougali B, Allaf K (2010) Instant controlled pressure drop extraction of lavandin essential oils: fundamentals and experimental studies. J Chromatogr A 1217(44): 6807–6815. https://doi.org/10.1016/j.chroma.2010.08.050

40. Socaci SA (2020) Food preservation and waste exploitation. IntechOpen, London. https://doi.org/10.5772/intechopen.78920

41. Iguedjtal T, Louka N, Allaf K (2008) Sorption isotherms of potato slices dried and texturized by controlled sudden decompression. J Food Eng 85(2):180–190. https://doi.org/10.1016/j.jfoodeng.2007.06.028

Index

Mohsen Gavahian (ed.), *Emerging Food Processing Technologies*,
Methods and Protocols in Food Science, https://doi.org/10.1007/978-1-0716-2136-3,

N

O

P

R

S

T

U

V

W

Y